AF361800

Applications of Artificial Intelligence in Medicine Practice

Applications of Artificial Intelligence in Medicine Practice

Editors

Kyungtae Kang
Junggab Son
Hyo-Joong Suh

MDPI • Basel • Beijing • Wuhan • Barcelona • Belgrade • Manchester • Tokyo • Cluj • Tianjin

Editors
Kyungtae Kang
Hanyang University
Korea

Junggab Son
Kennesaw State University
USA

Hyo-Joong Suh
The Catholic University of Korea
Korea

Editorial Office
MDPI
St. Alban-Anlage 66
4052 Basel, Switzerland

This is a reprint of articles from the Special Issue published online in the open access journal *Applied Sciences* (ISSN 2076-3417) (available at: https://www.mdpi.com/journal/applsci/special_issues/AI_medicine_practice).

For citation purposes, cite each article independently as indicated on the article page online and as indicated below:

LastName, A.A.; LastName, B.B.; LastName, C.C. Article Title. *Journal Name* **Year**, *Volume Number*, Page Range.

ISBN 978-3-0365-4423-6 (Hbk)
ISBN 978-3-0365-4424-3 (PDF)

Contents

About the Editors

Kyungtae Kang

Kyungtae Kang (Professor) received his B.S. degree in computer science and engineering, followed by M.S. and Ph.D. degrees in electrical engineering and computer science, from Seoul National University, Seoul, Korea, in 1999, 2001, and 2007, respectively. From 2008 to 2010, he was a postdoctoral research associate at the University of Illinois at Urbana-Champaign, IL, USA. In 2011, he joined the Department of Computer Science and Engineering at Hanyang University, where he is currently a tenured professor. His research interests lie primarily in systems, including operating systems, mobile systems, distributed systems, and real-time embedded systems. His recent research interest is in the interdisciplinary area of cyber-physical systems.

Junggab Son

Junggab Son (Assistant Professor) is currently an Assistant Professor in the Department of Computer Science, College of Computing and Software Engineering, Kennesaw State University (KSU), Marietta, GA, USA. He was a Limited-term Assistant Professor from January to May 2018 and was a research fellow/a part-time assistant professor from October 2016 to December 2017 in the Department of Computer Science, KSU. Before joining KSU, he was a postdoctoral research associate in the Department of Mathematics and Physics, North Carolina Central University, Durham, NC, USA from September 2014 to September 2016. He received his Ph. D. degree (Aug. 2014) and M.S. degree (Feb. 2011) in computer science and engineering from Hanyang University, Seoul, South Korea. He received his B.S. degree (Feb. 2009) in computer science and engineering from Hanyang University, Ansan, South Korea. His research interests include applied cryptography, security and privacy issues on significant applications, which includes cloud computing (Fog/Edge Computing), internet of things (Future Internet), vehicular ad hoc network, social network services, and bioinformatics.

Hyo-Joong Suh

Hyo-Joong Suh (Professor) is currently a professor at the School of Computer Science and Information Engineering, the Catholic University of Korea. He received his BS and MS degrees from Seoul National University in 1992 and 1994, respectively. He completed his PhD degree from the Department of Computer Engineering of Seoul National University in 2000. He is an expert in embedded and mobile systems with extensive experience in scalable computer and wireless/mobile systems. His research extends from memory hierarchy optimization during his MS and PhD researches to the prototyping of various mobile devices on several communication companies as a professional service. His current research interest focuses on human behavior computing with personal identification by using various sensors.

Preface to "Applications of Artificial Intelligence in Medicine Practice"

This book presents advanced research on AI theory and its applications in medicine, medically oriented human biology, and general healthcare from a variety of multidisciplinary perspectives. AI in biomedicine and clinical medicine, machine learning-based decision support, robotic surgery, data analytics and mining, laboratory information systems, and AI in medical education are among the themes covered. Following this are the results of a study on the "Convergence and Open Sharing System" Project, supported by the Ministry of Education and National Research Foundation of Korea.

Kyungtae Kang, Junggab Son, and Hyo-Joong Suh
Editors

 *applied sciences*

Editorial

Application of Artificial Intelligence in the Practice of Medicine

Hyo-Joong Suh [1], Junggab Son [2] and Kyungtae Kang [3,*

[1] School of Computer Science and Information Engineering, The Catholic University of Korea, Bucheon 14462, Korea; hjsuh@catholic.ac.kr
[2] Department of Computer Science, Kennesaw State University, Marietta, GA 30060, USA; json@kennesaw.edu
[3] Department of Artificial Intelligence, Hanyang University, Ansan 15588, Korea
* Correspondence: ktkang@hanyang.ac.kr

Citation: Suh, H.-J.; Son, J.; Kang, K. Application of Artificial Intelligence in the Practice of Medicine. *Appl. Sci.* **2022**, *12*, 4649. https://doi.org/ 10.3390/app12094649

Received: 24 April 2022
Accepted: 4 May 2022
Published: 6 May 2022

Publisher's Note: MDPI stays neutral with regard to jurisdictional claims in published maps and institutional affiliations.

1. Introduction

Advancements in artificial intelligence (AI) based on machine and deep learning are transforming certain medical disciplines. When combined with the rapid progress in high-performance computing, AI-based systems have enhanced the accuracy of diagnostics and the efficiency of therapeutics in many specializations. Advanced AI algorithms can extract features from a significant amount of healthcare data and then apply them to clinical practice. Furthermore, depending on feedback, the algorithm's accuracy is improved by its self-correcting abilities. Consequently, an AI-based healthcare support system can help physicians deliver optimal patient care by reducing diagnostic and therapeutic errors that unavoidably occur in human-based clinical practice [1]. In addition, such AI-based systems can extract meaningful information from a large patient population's data to draw real-time conclusions related to health risk alarms and health outcome projections.

According to experts, diverse healthcare sectors including chronic illness management and clinical decision-making can expect to be substantially impacted by AI. While AI algorithms are still in the early stages of deployment, they show promise in fields including radiology, pathology, ophthalmology, and cardiology [2]. Such progress poses interesting questions about whether AI will eventually displace clinicians, enhance their professional prospects, or some combination of both.

This Special Issue's objective is to advance research into a wide range of multidisciplinary perspectives on AI theory and its applications in medicine, medically oriented human biology, and general healthcare. The topics covered include (but are not limited to) AI in biomedicine and clinical medicine, machine learning-based decision support, robotic surgery, data analytics and mining, laboratory information systems, and AI in medical education. We stress the practical aspects of each study, emphasizing the importance of including a clinical evaluation of the utility and potential impact of the work.

2. Review of Issue Contents

This Special Issue presents ten original papers that cover the latest technologies and advances in the design of intelligent medical systems and applications. Moreover, each paper contributes to research that affords insights into the processing of medical data collected from patients.

Visual acuity (VA) measures the ability to distinguish the shapes and details of objects at a given distance. However, in some cases, such as unconsciousness or disease e.g., dementia, it may be impossible to measure VA using traditional chart-based methods. In [3], Kim et al. propose a machine-learning-based VA measurement method that determines VA from fundus images only. Three models, SVM, VGG-19, and EfficientNet-B7, were ensembled to predict categories. This is a precedent for applying artificial intelligence in medical practice to measure VA using fundus images.

Neuroimaging must often process a large amount of data with significantly fewer cases than the number of variables, which results in overmatching. To prevent this problem, Belenguer-Llorens et al. [4] propose a new dual Bayesian linear regression model with feature selection (DBL-FS) that effectively reduces the number of samples with high-dimensional features. This relies on including an automatic relevance determination prior (ARD) over the weight matrices, which automatically infers the features' relevance in the input feature space by assigning higher/lower relevance values when they contain more/fewer relevant features.

In addition, the DBL-FS Bayesian approach facilitated prior expert knowledge to guide the FS process and compensated for the limited number of samples available to train the model. The advantage of using DBL-FS allowed the detection and characterization of morphometric brain changes in a schizophrenic rodent model.

Image segmentation is used to analyze medical images quantitatively for diagnosis and treatment planning. This is because manual segmentation requires considerable expert effort and time. Ju et al. [5] propose a deep learning tool that easily creates training data to mitigate this inconvenience. This study was performed using two types of information: visual features and organ segment locations. The proposed model consists of two submodules: a feature encoder and a kernel function. The kernel function incorporates feature similarity density and Gaussian kernel density. The tool demonstrates competitive results when compared to state-of-the-art segmentation algorithms, such as UNet and DeepNetV3. The tool can be trained with minimal labeled data, uses anchor pixels from user interactions to segment organs easily, and refines the segmentation results by modifying the thresholds. Hofmann et al. [6] used machine learning to predict whether patients with schizophrenia exhibit aggressive behaviors. Up-sampling was used to process a small number of categories to balance the data, reduce variables using the random forest algorithm, and build machine learning models by including the logistic regression, trees, random forest, gradient boosting, k-nearest neighbor (KNN), support vector machines (SVM), and naive Bayes approaches. The performance of the SVM model was superior to the other machine learning algorithms. Negative behavior towards other patients was identified as the most indicative factor for distinguishing aggressive from non-aggressive patients. Its application may enable clinicians to identify high-risk patients at an early stage, modify their treatment accordingly, and prevent aggressive events during hospitalization.

Identifying the locations and extent of brain infarctions is essential for diagnosis and treatment. In general, deep learning requires large amounts of training data. To overcome this problem, Yoshida et al. [7] generated pseudo-patient images using CycleGAN, which performed image transformation without paired images. First, CycleGAN was used for data augmentation and to generate pseudo-cerebral infarction images from images of healthy specimens. Finally, U-Net was used to segment the cerebral infarction region using the CycleGAN-generated images. Regarding extraction accuracy, the U-Net-with-CycleGAN images showed an improvement over those of U-Net without CycleGAN, were efficient, and assisted in extracting the infarction area accurately while maintaining the detection rate.

STHarDNet [8] is a novel segmentation model for magnetic resonance imaging (MRI). In MRI segmentation, conventional approaches utilize U-Net models with encoder–decoder structures, segmentation models using vision transformers, or models that combine a vision transformer with an encoder–decoder model structure. However, conventional models are large with low computation speeds, and, in vision transformer models, the amount of computation sharply increases with the image size. To overcome these problems, the STHarDNet model is proposed, which combines Swin transformer blocks and a lightweight U-Net-type model that has a HarDNet block-based encoder–decoder structure. To maintain the features of the hierarchical transformer and shifted windows approach of the Swin transformer model, the Swin transformer is used in the first skip connection layer of the encoder, instead of in the encoder–decoder bottleneck.

STHarDNet improved the accuracy and speed of MRI image-based stroke diagnosis. In general, combined, the Swin transformer blocks and lightweight U-Net type model maintained the advantage of hierarchical feature extraction and demonstrated excellent segmentation performance. The Swin transformer restricts the computation of attention to each window, and this also maintains high calculation speeds.

The whole-slide image (WSI) is a digitized medical image. Processing WSIs to train neural networks is often intricate and labor-intensive. Neuner et al. [9] developed an open-source library dealing with recurrent tasks in the processing of WSIs and helped with the training and evaluation of neuronal networks for classification tasks. First, a large WSI is divided into multiple small tiles. Thereafter, the region of interest (ROI) is extracted using a filtering algorithm that stores each WSI's dimensions, ROI, and tile information. In addition, evaluations are available at each level while preserving the hierarchical structure. Neural network training continues using the fastai library, which applies filtered information for learning, reduces storage space, and increases the processing speed. This approach supplements the clinicopathological diagnoses of brain tumors.

Upper gastrointestinal endoscopy is widely performed to detect early gastric cancers (GCs). The automated detection of early GCs from endoscopic images involves an object detection model. However, the reduction of false positives involves challenges in the detected results. Teramoto et al. [10] propose an object detection model, U-Net R-CNN, based on a semantic segmentation technique that extracts target objects by performing local analysis on the images. The candidate regions were extracted using U-Net; however, many regions were over-detected in the detected candidate regions. Therefore, the candidate region was cut and input to the CNN to classify the candidate region as a GC or a false positive. Finally, the regions identified by the CNN were considered candidate regions. DenseNet169 was used as the convolutional neural network for box classification, which improved the detection performance compared with the previous method.

In [11] the authors verified that adversarial attacks were not negligible during open-source development. Open-source deep neural networks (DNNs) for medical imaging are significant in emergent situations, such as during the COVID-19 pandemic because they accelerate the development of high-performance DNN-based systems. The COVID-Net model, an open-source DNN model for detecting COVID-19 from chest X-ray images, is susceptible to backdoor attacks that modify DNN models and cause misclassification when a specific input trigger is added. The backdoor attacks are effective against models fine-tuned from the backdoored COVID-Net models, although non-targeted attacks are less successful. This indicates that the high-risk backdoored models can be spread by fine-tuning, thereby becoming a significant security threat. The findings show that protection must be emphasized during open-source development and in the practical application of DNNs for COVID-19 detection.

Finally, in [12], Calnares et al. present an automatic system for modeling clinical knowledge to follow a physician's reasoning during medical consultation. Instance-based learning was applied to provide suggestions for electronic medical records. A learning method was applied to determine the case types that best match the clinical scenarios of patients being evaluated according to an ad hoc similarity metric. A list of similar case types was suggested during evaluation whenever the physician modified the patient's information. The list of similar case types was updated when introducing or removing any clinical phase during medical consultation. This learning method can produce suggestions within a reasonable timeframe, even when processing large volumes of data. It is a novel tool that helps meet healthcare goals and reminds physicians to record essential data to fulfill care goals.

3. Conclusions

AI is a frontier where powerfully disruptive computer science advances have the potential to transform fundamentally the practice of medicine and healthcare delivery. It is profoundly changing the traditional model of medicine and significantly improving the level of medical services to assure various aspects of human health. Ever broader prospects are anticipated for the development of medical AI. Based on this trend, this special volume presents new and innovative research addressing some of the many scientific challenges associated with applying AI in medicine. We emphasize the need for a better understanding of AI's ongoing incorporation into routine medical practice. As such, the studies in this volume provide valuable perspectives on AI's future in healthcare, describe a roadmap for building effective, reliable, and safe approaches to AI in medicine, and discuss potential directions for developing AI-augmented healthcare systems.

Author Contributions: Conceptualization, H.-J.S., J.S. and K.K.; methodology, H.-J.S. and K.K.; validation, J.S.; investigation, K.K.; writing—original draft preparation, K.K.; writing—review and editing, H.-J.S.; supervision, J.S. and K.K.; funding acquisition, H.-J.S. and K.K. All authors have read and agreed to the published version of the manuscript.

Funding: This research was supported by the Catholic University of Korea, Research Fund, 2021. This research was also supported by the Institute of Information & Communications Technology Planning & Evaluation (IITP) grant funded by the Korean government (Ministry of Science and ICT) (No. 2020-0-01343, Artificial Intelligence Convergence Research Center (Hanyang University ERICA) and 2021-0-01547, High-Potential Individuals Global Training Program).

Institutional Review Board Statement: Not applicable.

Informed Consent Statement: Not applicable.

Acknowledgments: This Issue would not have been possible without the help of a variety of talented authors, professional reviewers. First, we express our gratitude to the authors for their excellent contributions to this special issue on application of artificial intelligence in the practice of medicine. We are also grateful to all the reviewers for their time and effort in examining these papers, and for their valuable comments and constructive suggestions. We hope that this special issue will serve as a valuable reference for academicians, scientists, engineers, and practitioners working toward the application of artificial intelligence in the practice of medicine.

Conflicts of Interest: The authors declare no conflict of interest.

References

1. Buch, V.; Ahmed, I.; Maruthappu, M. Artificial Intelligence in Medicine: Current trends and future possibilities. *Br. J. Gen. Pract.* **2018**, *68*, 143–144. [CrossRef] [PubMed]
2. Briganti, G.; Moine, O. Artificial Intelligence in Medicine: Today and Tomorrow. *Front. Med.* **2020**, *7*, 27. [CrossRef] [PubMed]
3. Kim, H.J.; Jo, E.; Ryu, S.; Nam, S.; Song, S.; Han, Y.S.; Kang, T.S.; Lee, W.; Lee, S.; Kim, K.H.; et al. A Deep Learning Ensemble Method to Visual Acuity Measurement Using Fundus Images. *Appl. Sci.* **2022**, *12*, 3190. [CrossRef]
4. Belenguer-Llorens, A.; Sevilla-Salcedo, C.; Desco, M.; Soto-Montenegro, M.L.; Gómez-Verdejo, V. A Novel Bayesian Linear Regression Model for the Analysis of Neuroimaging Data. *Appl. Sci.* **2022**, *12*, 2571. [CrossRef]
5. Ju, M.; Lee, M.; Lee, J.; Yang, J.; Yoon, S.; Kim, Y. All You Need Is a Few Dots to Label CT Images for Organ Segmentation. *Appl. Sci.* **2022**, *12*, 1328. [CrossRef]
6. Hofmann, L.A.; Lau, S.; Kirchebner, J. Advantages of Machine Learning in Forensic Psychiatric Research—Uncovering the Complexities of Aggressive Behavior in Schizophrenia. *Appl. Sci.* **2022**, *12*, 819. [CrossRef]
7. Yoshida, M.; Teramoto, A.; Kudo, K.; Matsumoto, S.; Saito, K.; Fujita, H. Automated Extraction of Cerebral Infarction Region in Head MR Image Using Pseudo Cerebral Infarction Image by CycleGAN. *Appl. Sci.* **2022**, *12*, 489. [CrossRef]
8. Gu, Y.; Piao, Z.; Yoo, S.J. STHarDNet: Swin Transformer with HarDNet for MRI Segmentation. *Appl. Sci.* **2022**, *12*, 468. [CrossRef]
9. Neuner, C.; Coras, R.; Blümcke, I.; Popp, A.; Schlaffer, S.M.; Wirries, A.; Buchfelder, M.; Jabari, S. A Whole-Slide Image Managing Library Based on Fastai for Deep Learning in the Context of Histopathology: Two Use-Cases Explained. *Appl. Sci.* **2022**, *12*, 13. [CrossRef]
10. Teramoto, A.; Shibata, T.; Yamada, H.; Hirooka, Y.; Saito, K.; Fujita, H. Automated Detection of Gastric Cancer by Retrospective Endoscopic Image Dataset Using U-Net R-CNN. *Appl. Sci.* **2021**, *11*, 11275. [CrossRef]
11. Matsuo, Y.; Takemoto, K. Backdoor Attacks to Deep Neural Network-Based System for COVID-19 Detection from Chest X-ray Images. *Appl. Sci.* **2021**, *11*, 9556. [CrossRef]
12. Galnares, M.; Nesmachnow, S.; Simini, F. Instance-Based Learning Following Physician Reasoning for Assistance during Medical Consultation. *Appl. Sci.* **2021**, *11*, 5886. [CrossRef]

Article

A Deep Learning Ensemble Method to Visual Acuity Measurement Using Fundus Images

Jin Hyun Kim [1], Eunah Jo [1], Seungjae Ryu [1], Sohee Nam [1], Somin Song [1], Yong Seop Han [2,*], Tae Seen Kang [2], Woongsup Lee [1,*], Seongjin Lee [1], Kyong Hoon Kim [3], Hyunju Choi [4] and Seunghwan Lee [4]

[1] Department of AI Convergence Engineering, Gyeongsang National University, Jinju 52828, Korea; jin.kim@gnu.ac.kr (J.H.K.); chosd4603@gnu.ac.kr (E.J.); ruru0213@gnu.ac.kr (S.R.); namsohei1@gnu.ac.kr (S.N.); csarah013044@gnu.ac.kr (S.S.); insight@gnu.ac.kr (S.L.)

[2] Department of Ophthalmology, Institute of Health Sciences, Gyeongsang National University College of Medicine, Gyeongsang National University Changwon Hospital, Jinju 52828, Korea; tskang85@naver.com

[3] School of Computer Science and Engineering, Kyungpook National University, Daegu 37224, Korea; kyong.kim@knu.ac.kr

[4] Deepnoid Inc., Seoul 08376, Korea; hjchoi@deepnoid.com (H.C.); nasa10@deepnoid.com (S.L.)

* Correspondence: medcabin@hanmail.net (Y.S.H.); wslee@gnu.ac.kr (W.L.)

Abstract: Visual acuity (VA) is a measure of the ability to distinguish shapes and details of objects at a given distance and is a measure of the spatial resolution of the visual system. Vision is one of the basic health indicators closely related to a person's quality of life. It is one of the first basic tests done when an eye disease develops. VA is usually measured by using a Snellen chart or E-chart from a specific distance. However, in some cases, such as the unconsciousness of patients or diseases, i.e., dementia, it can be impossible to measure the VA using such traditional chart-based methodologies. This paper provides a machine learning-based VA measurement methodology that determines VA only based on fundus images. In particular, the levels of VA, conventionally divided into 11 levels, are grouped into four classes and three machine learning algorithms, one SVM model and two CNN models, are combined into an ensemble method in order to predict the corresponding VA level from a fundus image. Based on a performance evaluation conducted using randomly selected 4000 fundus images, we confirm that our ensemble method can estimate with 82.4% of the average accuracy for four classes of VA levels, in which each class of Class 1 to Class 4 identifies the level of VA with 88.5%, 58.8%, 88%, and 94.3%, respectively. To the best of our knowledge, this is the first paper on VA measurements based on fundus images using deep machine learning.

Keywords: visual acuity; fundus images; machine learning; ophthalmology; deep learning; SVM

Citation: Kim, J.H.; Jo, E.; Ryu, S.; Nam, S.; Song, S.; Han, Y.S.; Kang, T.S.; Lee, W.; Lee, S.; Kim, K.H. et al. A Deep Learning Ensemble Method to Visual Acuity Measurement Using Fundus Images. *Appl. Sci.* **2022**, *12*, 3190. https://doi.org/10.3390/app12063190

Academic Editor: Andrés Márquez

Received: 24 January 2022
Accepted: 11 March 2022
Published: 21 March 2022

Publisher's Note: MDPI stays neutral with regard to jurisdictional claims in published maps and institutional affiliations.

1. Introduction

According to World Health Organization (WHO) statistics, the number of people with visual morbidity worldwide, as of 2020, is in excess of 299.1 million, of which 49.1 million is blind [1]. Visual sight is closely related to the quality of daily human life, such as safe walking, driving, and working; thus, regular eye health screening is essential to maintain eye health. Visual Acuity (VA) is a measure of the ability of the eye to distinguish shapes and the details of objects at a given distance. It is one of the essential indications of health. It is the most commonly used intuitive measure of the visual system's performance. The measurement of VA provides a baseline recording of VA, aids examination and diagnosis of eye disease or refractive errors, assesses any vision changes, and measures the outcomes of cataract or other surgery.

VA may be measured in various ways, depending on various conditions, such as illuminance. However, the measurement of VA needs to be consistent in order to detect any changes in vision. The general ways of VA measurement are (1) multi-letter Snuellen or E chart (2) plain occluder, card or tissue, (3) pinhole occluder, (4) touch or flashlight, or

(5) patient's documentation. The general procedure for VA measurement recommended by the US national library of medicine can be as follows [2]:

1. Position the patient, sitting or standing, at a distance of 6 m from the chart;
2. Test the eyes one at a time, at first without any spectacles (if worn);
3. Ask the patient to cover one eye with a plain occluder, card or tissue;
4. Ask the patient to read from the top of the chart and from left to right;
5. Record the VA for each eye in the patient's notes, stating whether it is with or without correction (spectacles). For example, Right VA = 0.1 with correction, Left VA = 0.2 with correction.
6. ...

In this procedure, the communication between VA examiner and examinee is essential to measure VA. However, it is not suitable or impossible to use the classical ways of measuring VA in the following occasions:

- When the patient is unable to use the measurement tool due to mobility difficulties;
- When the patient is unconscious state or lack of cooperation during the evaluation;
- When malingering should be strongly suspected;
- When an infant or a very young patient requires the measurement of VA.

In particular, continuous visual acuity measurement is necessary to secure the quality of life after regaining consciousness for patients who remain unconscious for a long time. However, since the existing method of measuring vision requires the patient to have a conversation with a tester, it is impossible to measure the vision of a prolonged unconscious. In addition, in a situation, such as the recent COVID-19 pandemic, as social isolation is prolonged, patients feel that it is harder to visit hospitals than before. For this reason, it becomes challenging to manage people's visual sight via the traditional method of measuring eyesight.

This paper provides a vision measurement method using deep learning-based ensemble methodology using fundus images. In this paper, we would overcome the following two problems:

- How can we measure the VA from an examinee who cannot communicate with the VA examiner or tries to present an incorrect VA value?
- How can we achieve a more accurate classifier when a dataset is fairly biased to certain classes in terms of the number of sample data?

Fundus photography involves photographing the rear of an eye, which is also known as the fundus. It is a photo image most popularly used in examining more than 38 types of eye diseases, such as age-related macular degeneration, neoplasm of choroid, choriorretinal inflammation or scars, glaucoma, retinal detachment and defects, and so on. Fundus imaging has been advanced to decrease preventable visual morbidity by allowing easy and timely fundus screening. In particular, the usability and portability of fundus screening have been continuously advanced for the last two decades. Furthermore, recently, there have been significant technological advances that have radicalized retinal photography. Improvements in telecommunications and smartphones are two remarkable breakthroughs that have made ophthalmic screening in remote areas a realizable possibility [3].

We address the above first problem with the high availability of fundus images. We would estimate the VA by capturing a fundus image from a VA examinee and using a VA classifier based on a deep machine learning technique. In this paper, 11 levels from 0.0 to 1.0 (step by 0.1) of VA levels are grouped into four classes according to ophthalmologist doctor's needs.

To tackle the second problem, we adopt an ensemble approach consisting of three machine learning models. In the medical field, it is very difficult to obtain a balanced size of a medical dataset because the dataset of normal cases is much larger than those of abnormal cases. The dataset of VA measurement results has the same issue; the cases of a lower VA level, in reality, are much less than those of a higher VA level. For this reason, it is difficult to adopt a classical CNN model for such unbalanced datasets of VA levels in a classical way.

In our ensemble approach, three machine learning models and techniques are combined to the classification of VA level groups or VA levels using their best classification performance.

The contributions of this paper are to

- to present a deep-learning-based VA measurement approach using fundus images;
- to demonstrate the feasibility and effectiveness of an ensemble approach to overcome the difficulties of obtaining datasets with a balanced size, and
- to present a VA measurement alternative for the examinee who is not easy or has no possible way to communicate with the VA examiner.

To the best of our knowledge, this is the first paper on the VA measurement based on fundus images using machine learning.

This paper is organized as follows: In Section 2, we present related work relevant to this work. In Section 3, we present a simple description on fundus images and the datasets consisting of fundus image and VA measurement data that we obtain from hospitals. Section 4 presents the main idea of our approach, a deep learning-based ensemble methodology for VA Measurement. We discuss the reason why an ensemble method is appropriate for our work and individual machine learning techniques comprising the ensemble method. In Section 5, we present the validation results of our proposed 4-Class VA classifier based on fundus images with various metrics of machine learning performance evaluation and a comparison of our ensemble method against the VGG-19-based CNN model. Finally, we conclude this paper in Section 6.

2. Related Work

Colenbrander [4] discusses the classical methods for VA measurement. Recently, some approaches to VA measurement have presented using various tools and smartphones [5–7]. Recently, ML using DNN has been actively used to diagnose, predict, and suggest medical treatment methods [8–12]. ML using DNN is also being actively used in ophthalmology [13–18]. Closely related to this study, there are some VA measurements using DNN [18–21]. For instance, Alexeeff et al. [21] develop a prediction model of final corrected distance visual acuity (CDVA) after cataract surgery, using machine learning algorithms based on electronic health record data. The fundus image is a universal and most actively used ophthalmic image for the diagnosis of various ophthalmic diseases. For this reason, recently, ML-based AI using this fundus image as training data are being actively studied for classification and prediction of eye diseases, such as diabetic retinopathy, glaucoma, and age-related macular degeneration [22–27]. However, to our best knowledge, no previous work presents ML for VA measurement, using fundus images and VA measurements of personals.

3. Datasets: Fundus Images and Vision Measurements

In this study, a vision acuity classification model is implemented based on personal's vision data and the relevant fundus images.

Fundus photography involves taking pictures of the back of the eye, also known as fundus. A special fundus camera consisting of a complex microscope attached to a flash-enabled camera is used for fundus photography. The main structures that can be visualized in fundus photography are the macula, the optic disc, and mid-peripheral retina with retinal vessels.

In Figure 1, the retina is the innermost layer of light-sensitive tissue in most vertebrates and some mollusks. The optics of the eye make the visual world concentrated on the retina into a two-dimensional image, and the retina converts the image into an electrical nerve stimulus to the brain to create a visual perception. The retina functions similarly to the camera's film or image sensor. The optic disc is the point where the axons of retinal ganglion cells converge and leave the eye. The optic disc in the normal human eye carries 1–1.2 million afferent nerve fibers from the eye to the brain.The optic disc is also the entry point for the major blood vessels that supply blood to the retina. The oval yellow area surrounds the fovea near the center of the retina of the eye, the area of sharpest vision. The human macula is about 5.5 mm (0.22 inches) in diameter. The macula of the human eye is where light is focused by the structures in front of the eye (cornea and lens). The photoreceptor cells in the macula are connected to nerve fibers and transmit visual information to the brain.

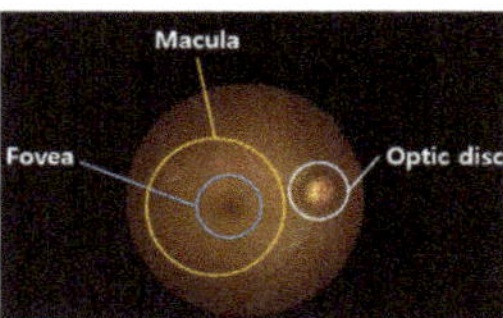

Figure 1. A normal fundus photograph of a right eye.

3.1. Datasets: Fundus Images and Patient's Vision Data

The vision chart data and fundus images of patients are obtained from 79,798 patients from February 2016 to January 2021 at the Department of Ophthalmology at Gyeongsang National University Changwon Hospital. The procedures used in this study followed the principles of the Declaration of Helsinki. The requirement for obtaining informed patient consent was waived by the institutional review board of Gyeongsang National University, Changwon Hospital (GNUCH 2021-05-007) due to the retrospective nature of the study.

The fundus images we used in this study are acquired in BMP files by an automation program in AutoIt. A fundus image is selected after matching the personal ID of a fundus image to a personal integrated vision information record. The original data used in this study are anonymized before its use. In this study, retrieving personal vision information necessary for machine learning is conducted in two stages: coupling fundus images and personal vision information and pre-processing of fundus images. In the first stage, we extract the vision acuity information from the medical charts of 79,798 patients with the keywords 'VA (Vision Acuity)', 'BCVA (Best Corrected VA)', and 'CVA (Corrected VA)' and reshape, for our purpose, personal vision datasets of 60,021 visual acuity information, of which each has a corresponding fundus image.

Initially, we have a total of 102,237 fundus images coupled with individual personal VA measurements. We use the personal id and the date of a funds image taken when coupling a fundus image and personal vision data. Ultimately, we obtained 79,800 images by this matching. Furthermore, 55,152 fundus images of them are used as data sets for machine learning.

We abstract the classical 11 levels of VA measurements (0.0–1.0, step by 0.1) into four groups as shown in Table 1. Class 1 consists of 2501 images with visual acuity of 0.0 to 0.05, Class 2 consists of 3972 images with 0.1, 0.15, and 0.2, Class 3 consists of 16,104 images with 0.3 to 0.7, and Class 4 consists of 32,575 images with 0.8 to 1.0. Table 1 shows the characteristics of fundus image findings for each level of VA. In addition to fundus image findings, visual acuity is dependent on optical and neural factors such as the sharpness of the retinal image within the eye, the function of the retina, and the interpretative function of the brain.

Table 2 shows three fundus images for each VA level and representative findings of each image. It will help to understand the characteristics of each class.

Table 1. Our VA classification.

Conventional VA Class	New VA Classes	Features
0.0–0.05	Class 1	1. Macular pigmentation and depigmentation findings 2. Macular bleeding and ischemia findings 3. Severe Peripapillary Atrophy orand Optic Nerve Atrophy 4. Tortuosity and abnormal findings of blood vessels near the macula 5. Overall cloudy fundus picture 6. Partially poorly observed fundus picture

Table 1. *Cont.*

Conventional VA Class	New VA Classes	Features
0.1, 0.15, 0.2	Class 2	1. Fundus findings similar to Class 1, 2. Fundus findings less cloudy than Class 1, 3. Less severe macular pigmentation and depigmentation and abnormal findings than Class 1.
0.3–0.7	Class 3	1. Fundus findings similar to Class 4, 2. Fundus findings that are generally more cloudy than Class 4, 3. Fundus findings that are partially cloudy than Class 4.
0.8–1.0	Class 4	1. Normal macula, 2. No dyspigmentation and no bleeding, 3. Normal optic disc.

Table 2. Examples of collected fundus images of Classes 1–4 and their features.

VA Classes	Fundus Images and Features
Class 1	(a) Macular pigmentation and depigmentation findings, (b) Macular hemorrhage and ischemic findings, optic nerve peripapillary atrophy and optic nerve atrophy, (c) Tortuosity and abnormal findings of blood vessels near the macula. Overall opinion: Overall cloudy fundus picture, partially obscured fundus picture.
Class 2	(d) Macular rarely visible, (e) Less cloudy than Class, (f) Less severe macular pigmentation and depigmentation and abnormal findings than Class 1.
Class 3	(g) Cloudy around macular, (h) Generally cloudy than Class 4. (i) Partially cloudy than Class 4.

Table 2. *Cont.*

VA Classes	Fundus Images and Features

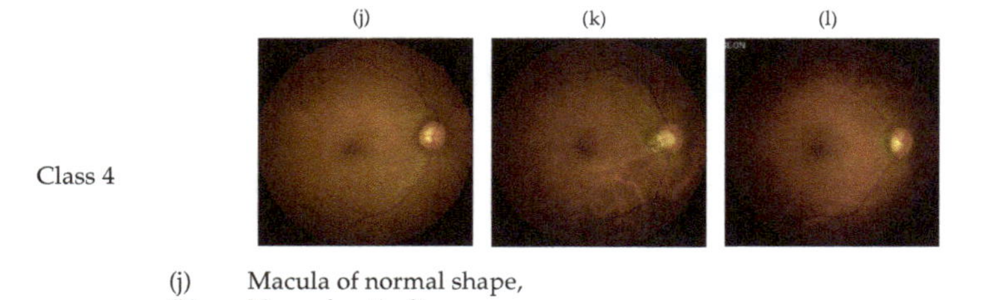

Class 4

(j)	Macula of normal shape,
(k)	Normal optic disc,
(l)	Abnormalities of the retina and blood vessels far from the macula.

3.2. Pre-Processing of Fundus Images

In the second step, fundus images are pre-processed with three filters and their combination, as shown in Figure 2, to augment and generalize fundus image data.

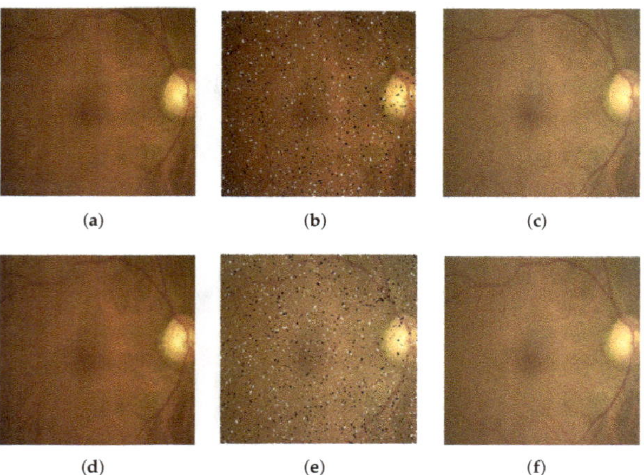

Figure 2. Pre-processing of fundus images. (**a**) Original; (**b**) Salt and pepper; (**c**) Gamma correction; (**d**) Remove noise; (**e**) Gamma correction + Salt and pepper; and (**f**) Gamma correction + Remove noise.

Table 3 summarizes the types and functions of pre-processing methods. Note that the pre-processing methods provided in this section are randomly applied to augment fundus images only for Classes 1 and 2.

Indeed, other pre-processing methods, such as shearing and shifting, may be helpful to improve the performance of the VA classifier. The image processing methods, such as shearing and shifting, that adjust the shape of images and the position of image features do not work effectively to improve the classification accuracy of our trained machines. For example, the shearing filter is not effective enough to improve the classification accuracy of VA measurement in our experiments. It seems that, when a CNN is trained, tweaking of the shape and shifting of the image location impair the shape of macular and optic nerve papilla that the human doctor observes carefully to check the health of the eye. Rotation of images is applied to augment fundus images from the datasets of Classes 1 and 2 which are much less than the other classes, and rescaling is limitedly applied fitting to our needs and purposes, such as transfer learning and SVM training.

Table 3. Pre-processing methods.

Pre-Processing	Features
Salt & Pepper	• It changes the pixel of an image to 0 or 255 with a certain probability. Instead of adding or reducing the value to the pixel value of the original image, the pixel exists in white or black with a specific probability.
Gamma correction	• It corrects brightness by changing pixel values if the image is too dark or bright. • The parameter is the gamma value γ. If $0 < \gamma < 1$, the image darkens; If $\gamma = 1$, the image changes the same as the original image; if $\gamma < 1$, the image brightens.
Remove noise	• The median filter, one of the ways to remove noise, sorts the values of surrounding pixels and changes the pixel value to the median value. It is mainly used to remove salt and pepper.

3.3. Data Augmentation

Table 4 shows the size of datasets of each VA level for our machine learning. For Classes 3 and 4, we do not augment or pre-process the datasets of Classes 3 and 4. The datasets of Classes 1 and 2 are much less than those of Classes 3 and 4. For the reason, we augment them in the following ways: first, we randomly select around 2500 from the dataset of Class 2 to make it balanced with the dataset of Class 1. Then, the two datasets of Classes 1 and 2 are augmented in the following way: the images of Classes 1 and 2 at the rate of 45% to 50% are randomly selected and rotated at $-10°$ to $10°$. Then, 25% to 30% images of the rotated images of Classes 1 and 2 are applied for the filtering methods in Section 3.2. Then, the datasets of Classes 1 and 2 are augmented the following way: the images of Classes 1 and 2 are selected from the original datasets randomly at the rate of 45% to 50% and rotated at $-10°$ to $10°$. Then, the filtering methods in Section 3.2 are applied for 25% to 30% images from the rotated images of Classes 1 and 2.

For all images of Classes 1 to 4, each image is cropped so that the main part of the macula and optic nerve papilla remains wholly highlighted, as shown in Figure 3, by completely removing the black part of each image. The original size of each image may not be identical to the others since they are captured in different fundus cameras. Thus, all images are resized in the size of 300×300.

Fundus images in all classes may be resized again for their individual methods when they are fed to CNN and SVM models for machine learning. For CNN models, the fundus images are resized into 244×244 fitting to the input image size of CNNs for transfer learning. For the SVM model, fundus images are resized into 32×32.

Table 4. The number of augmented datasets.

	Class 1	Class 2	Class 3	Class 4
Initial Number of fundus images	2501	3972	16,104	32,575
The number of augmented datasets	7109	7115	16,104	32,575

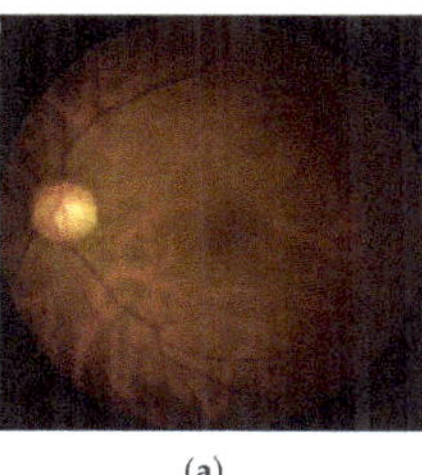
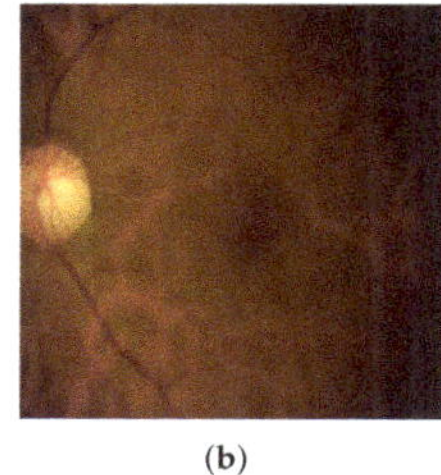

(a)   (b)

Figure 3. Before and after cropping fundus images. (**a**) before cropping; (**b**) after cropping.

4. Deep Learning-Based Ensemble Method for VA Measurement

This study presents the measurement of VA based on only fundus images. The conventional 11 classes of the VA (0.0–1.0, step by 0.1) are grouped into four classes. We devise an ensemble method consisting of three machine learning models, two CNN models and one SVM model, to overcome the quantity unbalance of the fundus images and improve the accuracy of the VA classification.

4.1. Ensemble Methods

This section discusses the rationale for the use of Ensemble methods consisting of three machine learning techniques: two deep neural network (DNN) models using transfer learning of Convolution Neural Network (CNN) techniques based on VGG19 [28] and EfficientNet-B7 [29], and Support Vector Machine (SVM).

We applied each technique of VGG19, EfficientNet-B7, and SVM for the original fundus images for 4-level VA classification, and the accuracy of each model could not exceed about 70%. Even after augmenting fundus images, the accuracy of the VA classifier with individual machine learning models could not be more than about 80%. To identify the cause of low accuracy of VA classification by each ML model, we analyzed confusion matrices in Figure 4 generated from the VGG-19 model.

Figure 4 shows the confusion matrix that the 4-level VA classifier based on VGG-19 returns when applied for each class of 11 VA levels. When the 4-level VA classifier is applied for Class-0.0 fundus images, the classification accuracy is 99% (Figure 4a). For Class-0.1 fundus images, the classifier makes 84% right decisions (Figure 4b). In the case of Class-0.4, only 53% of Class-0.3 fundus images are correctly classified (Figure 4e).

Based on our observation, we doubted that one of the main reasons for the misclassifications problem is the quantity unbalance between each class of the original datasets: The number of fundus images in Class 1 is 2501, 5% of the total datasets, while that of Class 4 is 32,575, accounting for 59% of the total datasets. To solve this problem, we propose an ensemble method, as shown in Figure 5. In this approach, a fundus image is classified in a hierarchical way, by different machine learning methods which perform the best classification performance at each classification step.

In the following, Classes 1 and 2 with a small number of fundus images are labeled as Class A, and Classes 3 and 4 with a large number of fundus images are labeled as Class B. Our method consists of three steps: In Step-1, a given fundus image is classified into either of Class A or Class B. In the following steps, the image classified into Class A is identified into either of Class 1 or Class 2 (Step-2-1), and the image in Class B is identified into either of Class 3 or Class 4 (Step-2-2).

We use three different ML models: VGG19-based CNN (implemented by Tensorflow), EfficientNet-B7-based CNN (implemented by PyTorch), and SVM-RBF-kernel [30]. Table 5 shows the VA classification accuracy of each ML model when those ML models perform each VA classification of 4-Class, Step-1, Step-2-1, and Step-2-2, respectively. We selected the highest accuracy technique for each stage and completed the entire model of fundus image's VA classification machine learning based on these results.

In our ensemble method, a fundus image is first identified as either Class A (Classes 1 and 2) or Class B (Classes 3 and 4) by VGG19-based-CNN. The fundus image classified as Class A (Classes 1 and 2) at Step 1 is identified, at Step-2-1, as either Class 1 or Class 2 by the SVM-RBF-Kernel. Similarly, the image identified as Class B (Classes 3 and 4) is further identified as either Class 3 or Class 4 by EfficientNet-B7-based CNN on Step-2-2.

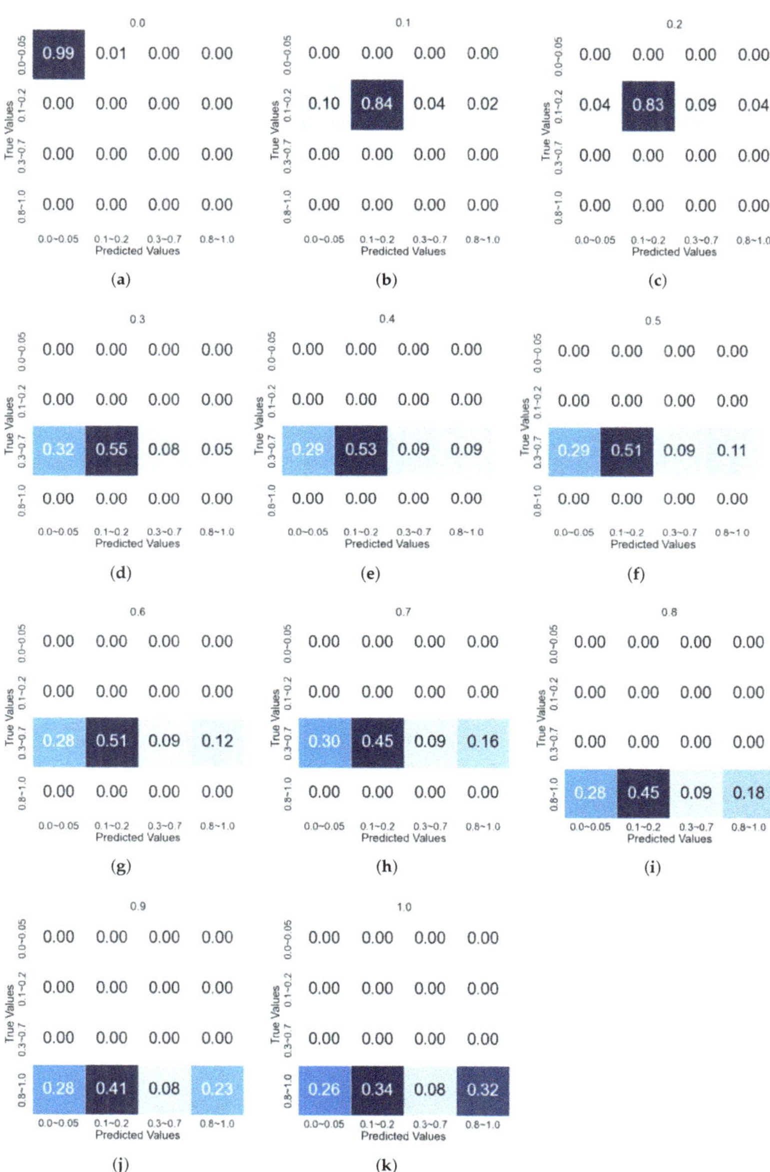

Figure 4. Confusion matrices from 4-Class VA Classifier's classification results for 11 classes of fundus images. (**a**) Class-0.0; (**b**) Class-0.1; (**c**) Class-0.2; (**d**) Class-0.3; (**e**) Class-0.4; (**f**) Class-0.5; (**g**) Class-0.6; (**h**) Class-0.7; (**i**) Class-0.8; (**j**) Class-0.9; (**k**) Class-1.0.

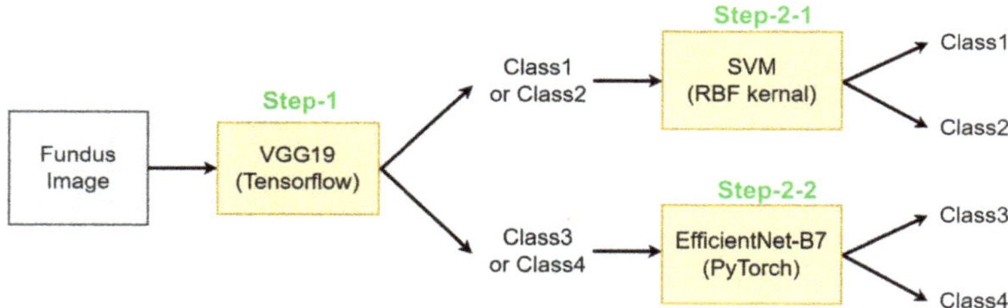

Figure 5. Proposed ensemble method.

Table 5. Classification accuracy when three machine learning models.

	4-Class Classification	Step-1	Step-2-1	Step-2-2
VGG19 (Tensorflow)	79%	**94%** [1]	79%	78%
EfficientNet-B7 (PyTorch)	78%	92%	78%	**79%**
SVM (RBF kernel)	77%	91%	**94%**	79%

[1] This accuracy number is the highest in this step (column).

4.2. ML Models and Classification Performances

Transfer learning is a machine learning technique that adopts the weight values of the pre-trained model to another machine learning. It is practical, i.e., fast and accurate when the number of training data are small because it reuses the weights of the pre-trained model. The pre-trained models VGG19 and EfficientNet-B7 used in this paper have a Convolutional Neural Networks (CNN) structure for image classification. The CNN consists of a convolutional base that extracts image features and a classifier that identifies the class of images based on the extracted features. In our transfer learning using VGG19 and EfficientNet-B7 models, the convolutional base is reused without any modification, and the classifier layers are built by ourselves.

4.2.1. ML Model for Step-1: VGG19-Based CNN

The VGG19 model was developed to study how the depth of the network affects the classification outcomes, such as accuracy and training speed. The convolutional base of the VGG19 consisting of the convolution and pooling is composed of 19 layers. All convolution layers are characterized by fewer parameters, using filters of 3×3. As a result, it can effectively extract features of images with small parameters, which leads to securing nonlinearity that can flexibly classify images. For Step-1, we build a VGG19 CNN model in which the convolutional base for feature extraction is based on VGG19 by transfer learning and the classification layers for the actual classification is built by ourselves, as shown in Figure 6. Table 6 shows the parameters we use to train DNN of VGG19 and EfficientNet-B7. We use a try-and-error approach to select the hyper-parameters after many experiments.

Figure 7a,b show the accuracy and loss changes of classifications during training and testing of the VGG19-based CNN model. Figure 7c shows a confusion matrix for 4000 images consisting of 1000 randomly selected from each class (Classes 1 to 4) of the original datasets, not augmented ones. The testing of VGG19-based CNN obtains 93% of the classification accuracy. From this confusion matrix, it is observed that some of Class A are identified as Class B. The reason might be inferred through the classification report in Figure 7d: The precision, which is the ratio of the data that is a true positive of the data determined to be positive, shows an accuracy of more than 80% in both Classes A and B. On the other hand, the recall, the ratio of positive data among the true positive data, is 60% for Class A compared to 98% for Class B. We believe that it is because of the data imbalance problem of the original datasets, thus the training of the VGG19 model is biased to Class B with a higher number of images. However, according to the f1-score in Figure 7d, which is the weight harmonic average of precision and recall and mainly used for unbalanced datasets, it can be inferred that the overall learning is well done because Class A shows the high accuracy of 70%.

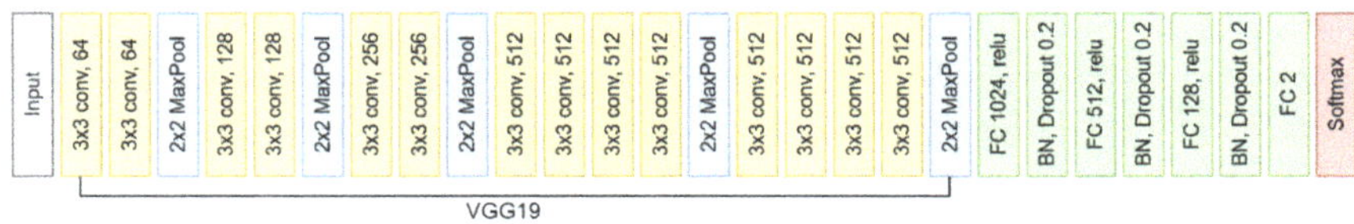

Figure 6. VGG19-based CNN model for VA classification.

Table 6. Learning parameters for VGG19 and EfficientNet-B7.

	Optimizer	Learning Rate	Batch Size	Epoch
VGG19 CNN	Adam	0.00002	128	500
EfficientNet-B7	Adam	0.00002	128	1000

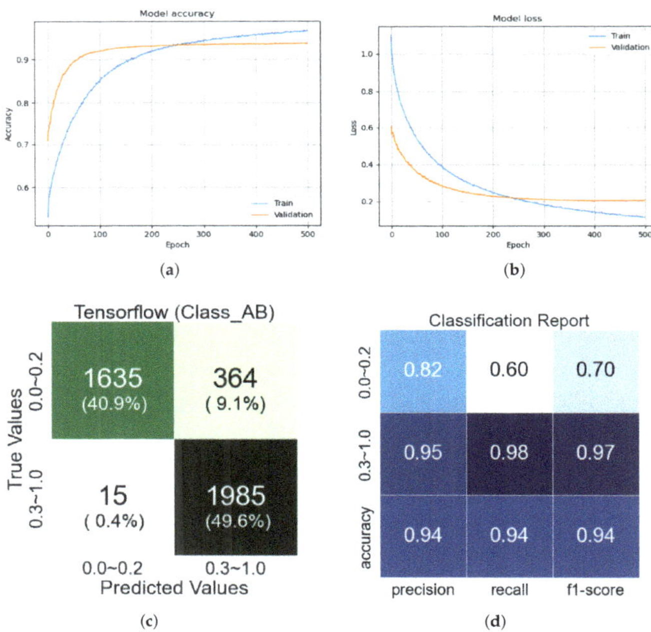

Figure 7. Training results of VGG19-based CNN for Step-1. (**a**) Accuracy; (**b**) Loss; (**c**) Confusion matrix; (**d**) Classification report.

4.2.2. ML Model for Step-2-1: SVM (RBF Kernel)

SVM is an algorithm based on statistical learning theory. It was initially devised to solve binary classification and regression analysis problems and extended for multiple classifications later. In addition, since the nonlinear separation between classes has been possible to solve using the notion of kernel method [31], it is popularly being used for data mining, artificial intelligence, prediction, and medical diagnosis.

In SVM, the learning data in a multidimensional space is expressed by:

$$\{x_i, y_i\}, i = 1, \ldots, n, y_i \in \{+1, -1\} \tag{1}$$

where x_i is a set of data and y_i is a label of x_i. For such learning data, several hyperplanes that separate the two classes can exist, but only one optimal hyperplane exists as shown in Figure 8a.

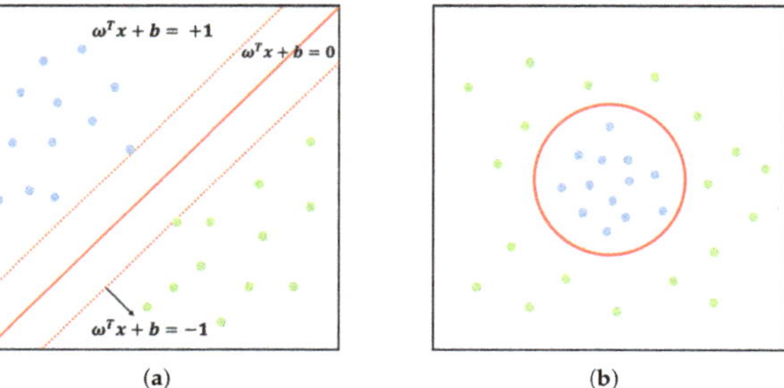

(a)   (b)

Figure 8. Hyperplanes in SVM. (**a**) Hard-margin; (**b**) soft-margin.

Such an optimal hyperplane maximizes the distance between the data closest to the hyperplane separated among each class of data. A hyperplane is defined by:

$$\omega^T \cdot x + b = 0 \tag{2}$$

If data are linearly separable as shown in Figure 8a, the hyperplane that separates the two classes can be defined by Equation (3):

$$\forall i, y_i(\omega^T \cdot x_i + b) - 1 \geq 0 \tag{3}$$

where ω is a vector of weight.

The training data for these two hyperplanes is called a support vector. In addition, since the margin between two hyperplanes must be maximized to obtain the hyperplanes of two classes, it becomes an optimization problem like the following objective of Equation (4) under Equation (3) as a constraint:

$$\min\{\frac{1}{2}||\omega||^2\} \tag{4}$$

In most cases, the data do not satisfy the above constraint because they are not linearly separable, as shown in Figure 8b. To solve this problem, the constraint is extended with a slack variable ξ representing the distance from the hyperplane to misplaced data, and the objective is extended with a penalty term c. As a result, we obtain an optimization problem as follows:

$$objective \quad : \quad \min\{\frac{1}{2}||\omega||^2 + c \sum_{i=1}^{n} \xi_i\} \tag{5}$$

$$s.t \quad : \quad y_i(\omega^T \cdot x_i + b) > 1 - \xi_i \tag{6}$$

For nonlinear data, a hyperplane of data having a nonlinear boundary can be obtained by data space transformation using $K(x_i, x_j)$ (Kernel function) for the features of data x_i and x_j. The representative $K(x_j, x_j)$ is defined by, respectively:

- Polynomial (Inhomogeneous): $K(\vec{x_i}, \vec{x_j}) = (\vec{x_i} \cdot \vec{x_j} + r)^d$;
- Radial basis function: $K(\vec{x_i}, \vec{x_j}) = exp(-\frac{||\vec{x_i} - \vec{x_j}||^2}{2\sigma^2})$;
- Sigmoid function: $K(\vec{x_i}, \vec{x_j}) = \tanh(\kappa \vec{x_i} \cdot \vec{x_j} + c)$ for some (not every) $\kappa > 0$ and $c < 0$, where κ is a gradient and c is a bias term (intercept).

Now, the SVM model using RBF kernel that we used for Step-2-1 is explained. First, we resize all the images from $300 \times 300 \times 3$ to $32 \times 32 \times 3$, as shown in Figure 9, which shows the data shape and features that SVM-RBF-Kernel would use to reason the class of a fundus image.

Figure 9. Fundus images compressed into $32 \times 32 \times 3$ for SVM (RBF kernel).

Using the SVM class in the scikit-learn and the resized dataset above, the SVM-RBF-Kernel model is trained by varying kernel functions, c that determines how much error the model tolerates, and γ that determines how flexible the hyperplane is set. In this study, we select the SVM model using the RBF kernel and train the model by setting γ to 0.1 and increasing c from 1 to 100 by 20. The confusion matrix from the testing of SVM-RBF-Kernel model is shown in Figure 10.

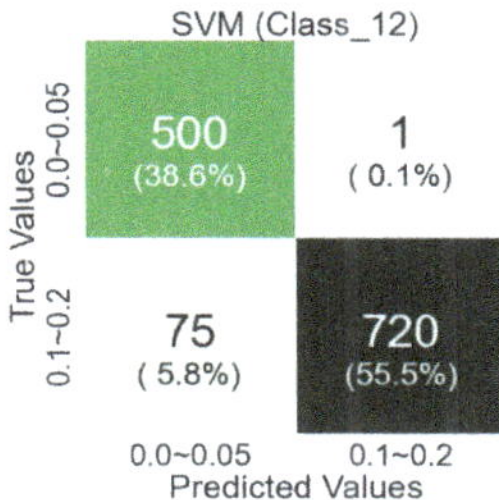

Figure 10. Confusion matrix of SVM-based classifier for Step-2-1.

In the confusion matrix from testing of an SVM-RBF-Kernel, the classification accuracy is about 58% for Class 1 and about 38% for Class 2, so the overall classification accuracy is about 96%. There are more Class 1 misclassified images than Class 2 misclassified images. The reason might be that there are similar abnormalities such as picture blurring or macular pigmentation and depigmentation in Class 2 as in Class 1.

4.2.3. ML Model for Step-2-2: EfficientNet-B7

EfficientNet is a state-of-the-art model with the best performance with a few parameters about image classification. The scaling-up method is often utilized to improve the performance of CNN. For scaling-up, one can increase the number of layers, the number of channels, or the input image's resolution. EfficientNet finds the optimal combination of the above three scaling-up methods through AutoML (Automated Machine Learning) by uniformly adjusting the three methods using compound coefficients.

EfficientNet is composed of MBConv structured as shown in Figure 11. The MBConv expands the channel through 1×1 convolution operation and performs a Depthwise convolution operation that performs a convolution operation on each image channel. Depthwise convolution performs a convolution operation with $k \times k$ kernel for each channel of images. Each channel operated by Depthwise convolution becomes a feature map. Each layer uses batch normalization and then goes through the Swish function as an activation function. The Swish function prevents the gradient value from being saturated near zero during learning, unlike the Sigmoid and Tanh functions. In addition, unlike Relu, the Swish is less sensitive to the initial value and learning rate. For an enormous negative value, the Swish function returns a value of 0, but it preserves the value to some extent for a small negative value. Squeeze and Excitation Layer is composed of Global Average Pooling-Fully Connected Layer-ReLU-Fully Connected Layer-Sigmoid. The two Fully Connected Layers prevent the number of parameters from increasing with a bottle-neck structure. Each channel's relative importance can be known through two Fully Connected Layers and nonlinear activation functions (ReLU, Sigmoid). The extracted map can be multiplied by a feature map that skips the Squeeze and Excitation Layer to highlight important features. Finally, the channel is reduced by the 1×1 convolution operation. For channels reduced to 1×1, using the activation function deletes sensitive information and it is less likely that sensitive information exists on other channels. Therefore, only batch normalization is used. In this way, the skip-connected input value is concatenated to the output value passed through multiple layers. This concatenation can preserve the previously learned information, learn additionally from it, and reduce memory usage.

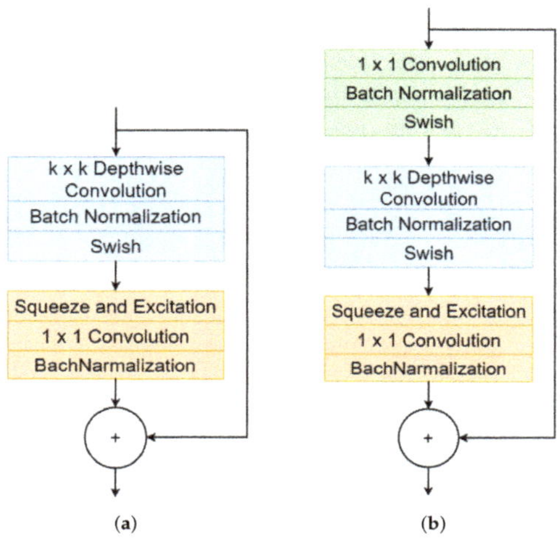

Figure 11. MBConv of EfficientNet [29]. (**a**) MBConv1; (**b**) MBConv6.

The EfficientNet-B7 model uses fewer parameters and provides the best performance among various EfficientNet models. As shown in Figure 12, the final model we apply for the VA classification in Step-2-2 is completed by transfer learning, in which the convolutional base for feature extraction is reused and the classifier part is directly built by ourselves.

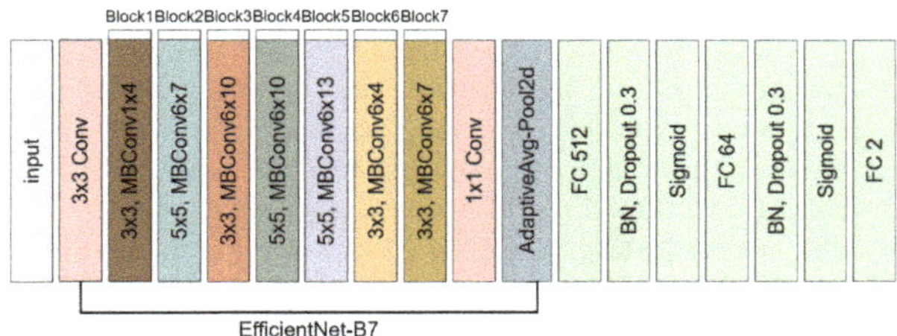

Figure 12. EfficientNet-B7 CNN model for VA classification in Step-2-2.

The parameters we used to train CNN based on EfficientNet-B7 are also given in Table 6.

In Figure 13a, the classification accuracy of testing data are about 79%, and in Figure 13b, the loss of testing data continues to drop, but the change of the loss value is insignificant after epoch 300. The confusion matrix in Figure 13c is computed with 20% of each class data randomly selected from the original datasets. In the classification report in Figure 13d, the f1-score of Class 3 (Class 0.3–0.7) is 0.51 and that of Class 4 (Class 0.8–1.0) is 0.79. The total f1-score is 0.70. The training of EfficientNet-B7-based CNN seems biased to Class 4, due to the imbalance of the data quantity, but the overall f1-score of the EfficientNet-B7-based CNN shows that the training of the CNN model is well done.

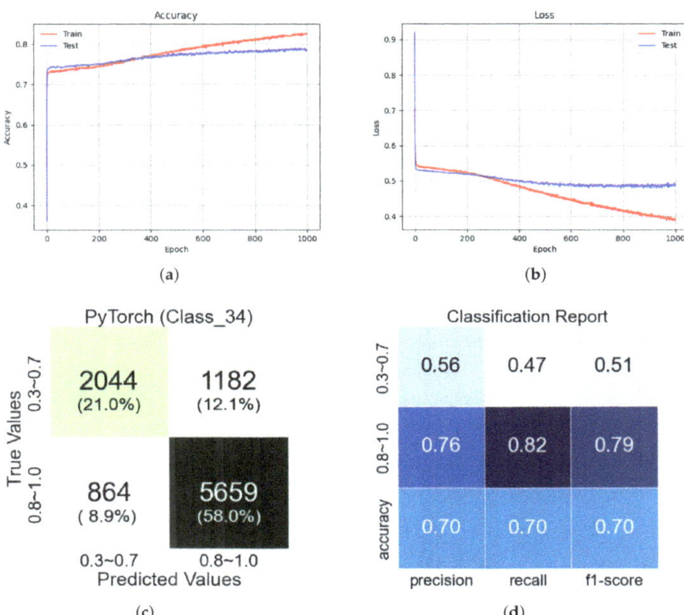

Figure 13. Training results of EfficientNet-B7-based CNN for Step-2-2. (**a**) Accuarcy; (**b**) Loss; (**c**) Confusion matrix; (**d**) Classification report.

5. Results

In this section, we present the experiment results where the 4-Class VA classifier in Figure 5 is applied for the validation datasets of the patient's fundus image and VA information. To figure out the accuracy of the overall model, we selected 1000 fundus images with patient VA information for each class from the original datasets and conducted a classification experiment. The selected fundus images are not pre-processed with any filter.

5.1. Validation of Classification of VGG19-Based CNN for Step-1

First, the VA classifier for Step-1 where Classes A and B are identified was tested by using VGG19-based CNN, and the confusion matrix is as shown in Figure 14.

The classification between Classes A and B is performed by VGG19-based CNN, of which the classification accuracy is around 94% as shown in Table 5. Our 4-Class VA classifier at Step-1 adopts the VGG19-based CNN, thus it cannot outperform the classification accuracy of VGG19-based CNN for VA classification.

In Figure 14a, the images in Class A are more misclassified than those in Class B. It is in accordance with the fact that the recall of Class A is 60% in Figure 7d. We investigated the probability value of Softmax of images misclassified in the confusion matrix and the characteristics of these images. In Table 2, the images in Class A look generally blurred and partially poorly observed. On the other hand, the images in Class B look relatively cleaner than Class A, and the macula and optic disc are clearly observable. However, the misclassified images of Class A in this experiment have the more observable optic disc and macula images than the images of Class A Table 2.

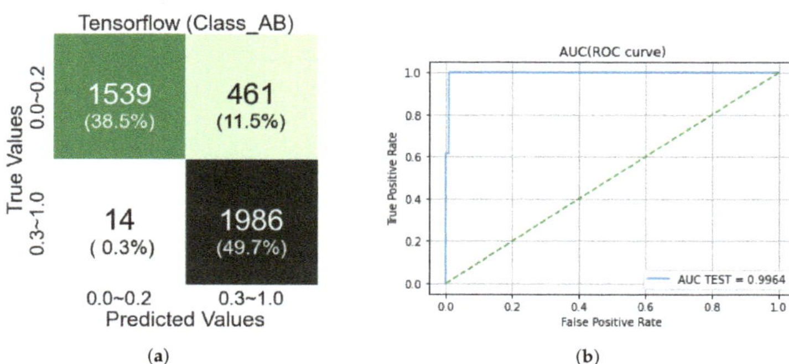

Figure 14. Verification results of 4-Class VA Classifier. (a) Confusion matrix; (b) Classification report.

Figure 15 shows some misclassification examples where the fundus images in Class 1 or Class 2 of Class A are misclassified as Class B. Note that the decision probability of Softmax of the fundus images in the first line is both close to 50%, and that means that they can likely be misclassified. Furthermore, the macular in fundus images in the second line look cleaner, in contrast to those in Table 2.

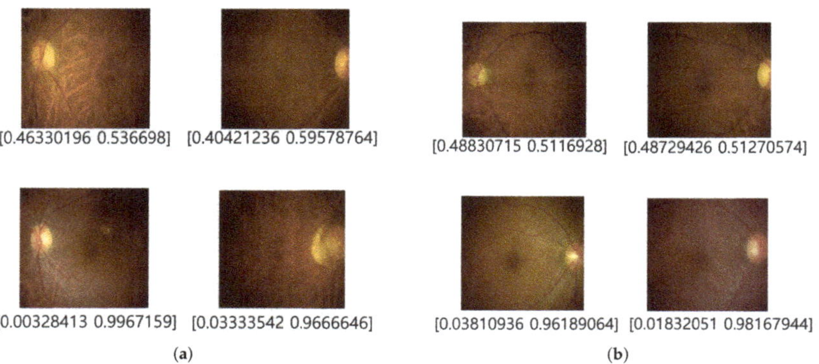

Figure 15. Examples of misclassification of Class A and the probabilities of Softmax: The pair of numbers in parenthesis are the probability values from Softmax. The first value is the probability that the image is in Class A and the second value is the probability that the image is in Class B. (a) Class 1 of Class A; (b) Class 2 of Class A.

For the misclassification cases of Class B, the misclassified images in Class B look very blurred or macularly unclear, as shown in Figure 16.

5.2. Validation of Classification of SVM-RBF-Kernel ML for Step-2-1

The fundus images of Class A identified in Step-1 by VGG19-based CNN are checked if it is in either Class 1 (0.0–0.05) or Class 2 (0.1–0.2) by using the SVM-RBF-Kernel.

Figure 17 shows a confusion matrix, and AUC score and ROC curve. In the confusion matrix, the classification accuracy for Classes 1 and 2 is about 58% and 38%, respectively, and the overall classification accuracy is about 96%.

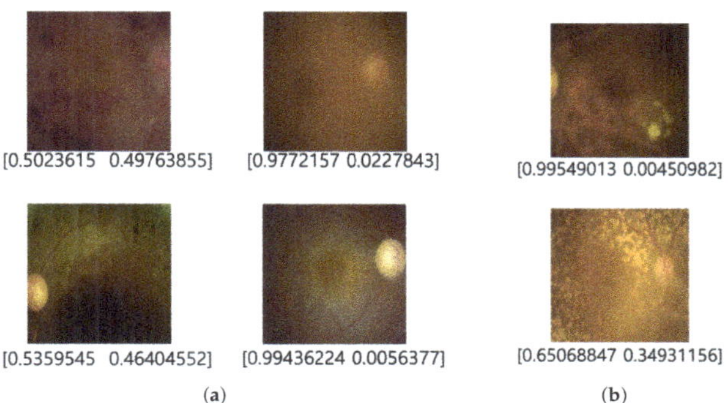

[0.5023615 0.49763855]   [0.9772157 0.0227843]   [0.99549013 0.00450982]

[0.5359545 0.46404552]   [0.99436224 0.0056377]   [0.65068847 0.34931156]

(a)   (b)

Figure 16. Examples of misclassification of Class B and the probabilities of Softmax: The pair of numbers in parenthesis are the probability values from Softmax. (**a**) Class 3 of Class B; (**b**) Class 4 of Class B.

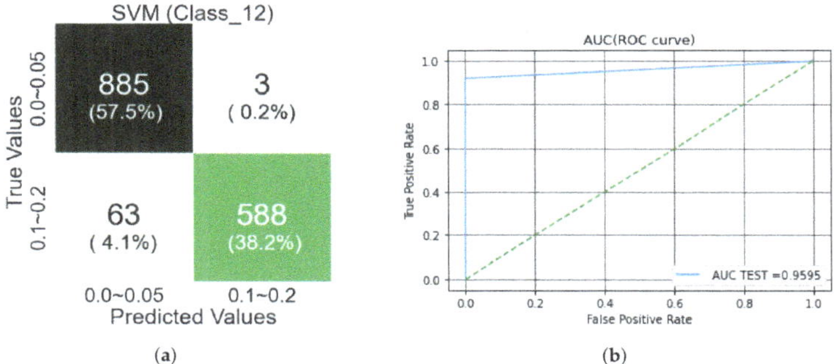

(a)   (b)

Figure 17. Validation results of SVM-RBF-Kernel classification for Step-2-1. (**a**) Confusion matrix; (**b**) AUC (ROC) report.

Tbltbl:fp-cls-features shows the images that should be classified as Class 1 but misclassified as Class 2 or vice versa: The misclassified fundus images of Class 1 (0.0–0.05) are cloudy, and each part of the fundus images is not easily identified. In addition, abnormal findings such as pigmentation and depigmentation of the macula are shown. On the other hand, Class 2 has fewer hazy fundus images and fewer abnormal findings such as macular pigmentation and depigmentation than Class 1.

In this validation, as shown in Figure 18, the fundus images of Class 1 misclassified as Class 2 are relatively clearer and have fewer abnormalities, such as pigmentation and depigmentation in the macula, than images classified correctly. The fundus Images of Class 2 misclassified as Class 1 look more blurred than images classified correctly, and the abnormalities, such as pigmentation and depigmentation, in the macula, appear more severe.

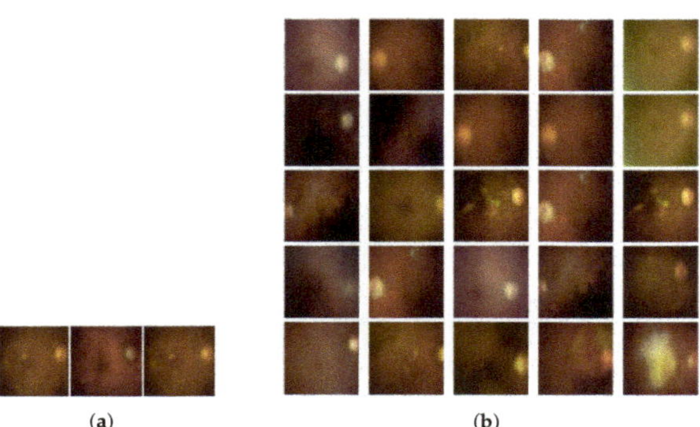

(a)   (b)

Figure 18. Examples of misclassification of Classes 1 and 2 by the SVM-RBF-Kernel. (**a**) Class 1; (**b**) Class 2.

5.3. Validation of Classification of EfficientNet-B7-Based CNN for Step-2-2

For the fundus images identified as Class B in the previous step, EfficientNet-B7-based CNN identifies the individual classes as either Class 1 or Class 2.

Figure 19 shows a confusion matrix and AUC score and ROC curve from the validation of the classification by EfficientNet-B7-based CNN in Step-2-2. In the confusion matrix, the classification accuracy for Classes 3 and 4 is about 44% and 48%, respectively, so the overall classification accuracy is 92%.

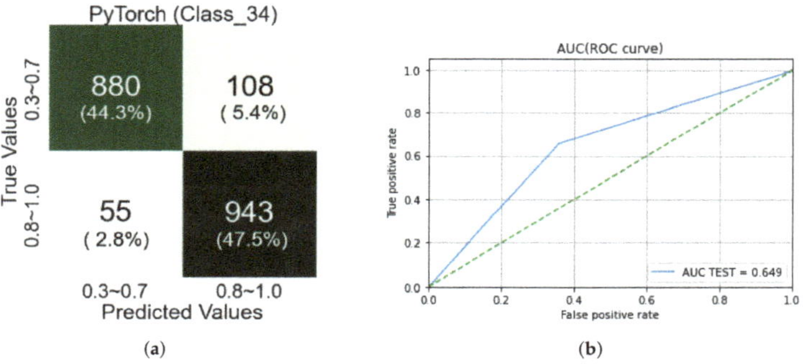

(a)   (b)

Figure 19. Validation results of EfficientNet-B7-based CNN classification for Step-2-2. (**a**) Confusion matrix; (**b**) AUC (ROC) report.

In Table 2, the optic disc and macula of the fundus images in Class 3 look clean and have no pigment abnormality or bleeding. On the other hand, the fundus images in Class 3 are an overall blurry image compared to Class 4. In the misclassification case of Class 3, the fundus images, as shown in Figure 20a, have the optic disc and macula in clearer. Meanwhile, in the misclassification case of Class 4, the optic disc and macula in the fundus image are not clearly observed, and a lot of blurred images are observed.

The conclusion about the classification accuracy of the 4-Class VA Classifier using fundus images is as follows: We randomly selected 1000 for each class and tested the classifier with a total of 4000 fundus images and the relevant patient's VA information. Figure 21 is a confusion matrix for the classification accuracy of the entire model based on the ensemble method. It combines Figures 14a, 17a and 19a and summarizes the classification performance of the 4-Class VA classifier. This confusion matrix consists of two types of quadrants: big quadrants and small quadrants. The big quadrants include the classification accuracy rate for Classes A and B performed at Step-1. The small quadrants include the classification accuracy rate for Classes 1–4 performed at Step-2. Notice the numbers on the diagonal in this confusion matrix, starting at the top left and flowing down to the right. These numbers are the classification accuracy for 1000 fundus images from each class from Class 1 to Class 4. It says that the classification accuracy of our approach to VA measurement based on fundus images are 88.5%, 58.8%, 88.0% and 94.3% for each classification of Class 1 to Class 4, respectively. We can say that the classification accuracy of the 4-Class VA classifier is 82.4% on average.

Table 7 shows the comparison between the performance of VA classifiers based on our ensemble method and VGG-19 in terms of four aspects: the overall average accuracy, each class accuracy, sensitivity, and specificity. The reason why VGG-19 is selected to compare against our ensemble method is that it shows the best performance of VA classification as shown in Table 5. It shows that our ensemble method outperforms the VGG-19 VA classifier in the overall accuracy, but the VGG-19 VA classifier shows higher accuracy in VA- classification for Class-2 than our ensemble method. From the aspects of sensitivity and specificity, they are not comparable because one of them does not outperform the other in all classes.

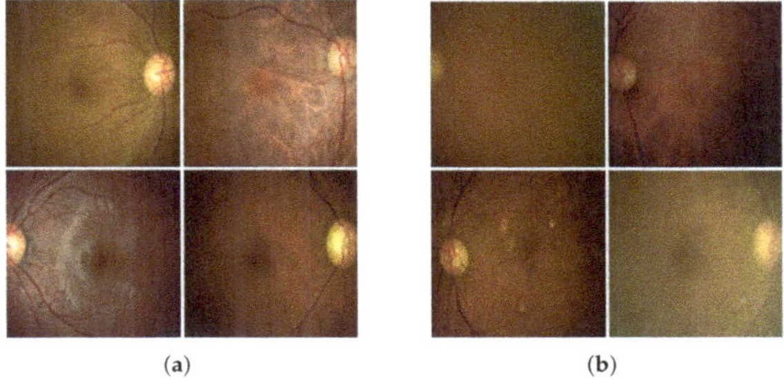

Figure 20. Examples of misclassification of Classes 3 and 4 by EfficientNet-B7. (**a**) Class 3; (**b**) Class 4.

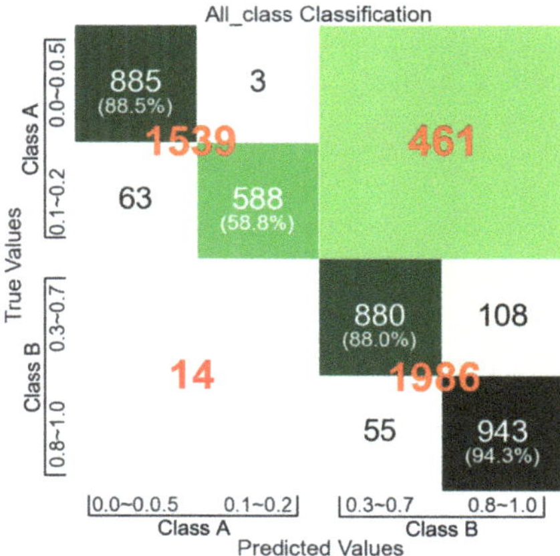

Figure 21. Overall accuracy report of 4-Class VA classifier based on our ensemble method.

Table 7. Comparison of VA classifiers' performance based on our ensemble method and VGG-19.

VA Classes	Ensemble Method				VGG-19			
	Average Accuracy	Class Accuracy	Sensitivity	Specificity	Average Accuracy	Class Accuracy	Sensitivity	Specificity
1	82.4%	88.5%	0.885	0.038	78%	0.89	0.988	0.0042
2		58.8%	0.588	0.121		0.8	0.832	0.06
3		88.0%	0.88	0.04		0.69	0.563	0.137
4		94.3%	0.943%	0.024		0.74	0.764	0.079

6. Conclusions

Visual sight is one of the most sensing capabilities of humans. Visual Acuity (VA) is a fundamental measure of the ability of the eye to distinguish shapes and the details of objects at a given distance. It is a primary indicator of eye health and the results of medical treatment for eye diseases. VA is typically measured using VA measuring tools, such as Snellen or E-Chart. In particular, communication with the tester is essential. However, it is not suitable or impossible to use the classical ways of measuring VA for patients under mobility difficulties, unconscious states, or lack of cooperation, and an infant or very young patient.

To solve those problems, we present an ensemble method based on machine learning based on fundus images and VA information of patients. Fundus photography is one of the most popularly used photo images for an eye examination and rarely needs the cooperation of patients to obtain the image. In our approach, 11 classes in classical VA measurement are abstracted into four classes, Classes 1–4, to overcome the discrepancy problem of fundus image data quantity for each of 11 classes.

In the ensemble method, the VA is measured in two steps: In the first step, Classes 1–2 and Classes 3–4 are classified as either Class A or Class B. In the second step, the fundus images in Class A is classified as either Class 1 or Class 2 and those in Class B is classified as either Class 3 or Class 4.

We use three different machine learning techniques for each classification: VGG-19-base CNN, EfficientNet-B7-based CNN, and SVM-RBF-Kernel. We evaluated the three techniques for each classification of individual steps and selected one of them that shows the best classification performance for each step. From our validation of the 4-Class VA classifier using 4000 fundus images from each of the four classes, we obtained 88.5%, 58.8%, 88 %, and 94.3% of classification accuracy for each level of four classes, respectively, and the classification accuracy of 82.4% on average.

To make our approach useful in practice, we have more challenges to overcome. For example, the density of the background pigmentation of the fundus oculi is dependent on race. We need to obtain more data from other countries and races to overcome this problem. In addition, the examinee's subjectivity in measuring vision acuity may degrade the collected data quality. In addition, the fundus image shows the functional status of the eye, thus measuring visual acuity with only fundus images have limitations since our vision depends on both the function of the eye and the function of the brain.

Author Contributions: Conceptualization, J.H.K., W.L. and Y.S.H.; methodology, J.H.K.; software, S.N., S.S., S.R. and E.J.; validation, S.N., S.S., S.R. and E.J.; investigation, S.N., S.S., S.R. and E.J.; data curation, T.S.K.; writing—original draft preparation, J.H.K., S.N., S.S., S.R. and E.J.; writing—review and editing, S.L. (Seongjin Lee) and K.H.K.; visualization, S.N., S.S., S.R. and E.J.; supervision, W.L. and Y.S.H.; project administration, J.H.K. and Y.S.H.; funding acquisition, J.H.K., H.C. and S.L. (Seunghwan Lee). All authors have read and agreed to the published version of the manuscript.

Funding: This research was funded by Regional Innovation Strategy (RIS) through the National Research Foundation of Korea(NRF) funded by the Ministry of Education (MOE)(2021RIS-003) and National Research Foundation of Korea(NRF) grant funded by the Korea government(MSIT) (No. 2020R1A2C1014855).

Institutional Review Board Statement: The protocol of this retrospective study was approved by the Institutional Review Board of Gyeongsang National University Changwon Hospital and followed the principles of the Declaration of Helsinki.

Informed Consent Statement: The requirement for obtaining informed patient consent was waived by the institutional review board (GNUCH 2021-05-007) due to the retrospective nature of the study.

Data Availability Statement: The fundus images of patients who visited Gyeongsang National University Changwon Hospital were obtained by an expert examiner using a digital retinal camera (CR-2; Canon, Tokyo, Japan).

Conflicts of Interest: The authors declare no conflict of interest.

References

1. Bourne, R.R.A.; Adelson, J.; Flaxman, S.; Briant, P.; Bottone, M.; Vos, T.; Naidoo, K.; Braithwaite, T.; Cicinelli, M.; Jonas, J.; et al. Global Prevalence of Blindness and Distance and Near Vision Impairment in 2020: Progress towards the Vision 2020 Targets and What the Future Holds. Available online: https://iovs.arvojournals.org/article.aspx?articleid=2767477 (accessed on 24 January 2022).
2. Marsden, J.; Stevens, S.; Ebri, A. How to Measure Distance Visual Acuity. Available online: https://www.ncbi.nlm.nih.gov/pmc/articles/PMC4069781/ (accessed on 24 January 2022).
3. Panwar, N.; Huang, P.; Lee, J.; Keane, P.A.; Chuan, T.S.; Richhariya, A.; Teoh, S.; Lim, T.H.; Agrawal, R. Fundus photography in the 21st century—A review of recent technological advances and their implications for worldwide healthcare. *Telemed.-Health* **2016**, *22*, 198–208. [CrossRef] [PubMed]
4. Colenbrander, A. The historical evolution of visual acuity measurement. *Vis. Impair. Res.* **2008**, *10*, 57–66. [CrossRef]
5. Bach, M. The Freiburg Visual Acuity Test-automatic measurement of visual acuity. *Optom. Vis. Sci.* **1996**, *73*, 49–53. [CrossRef] [PubMed]
6. Brady, C.J.; Eghrari, A.O.; Labrique, A.B. Smartphone-based visual acuity measurement for screening and clinical assessment. *JAMA* **2015**, *314*, 2682–2683. [CrossRef]
7. Tofigh, S.; Shortridge, E.; Elkeeb, A.; Godley, B. Effectiveness of a smartphone application for testing near visual acuity. *Eye* **2015**, *29*, 1464–1468. [CrossRef] [PubMed]
8. Kononenko, I. Machine learning for medical diagnosis: History, state of the art and perspective. *Artif. Intell. Med.* **2001**, *23*, 89–109. [CrossRef]
9. Foster, K.R.; Koprowski, R.; Skufca, J.D. Machine learning, medical diagnosis, and biomedical engineering research-commentary. *Biomed. Eng. Online* **2014**, *13*, 1–9. [CrossRef]
10. Erickson, B.J.; Korfiatis, P.; Akkus, Z.; Kline, T.L. Machine learning for medical imaging. *Radiographics* **2017**, *37*, 505–515. [CrossRef]
11. Willemink, M.J.; Koszek, W.A.; Hardell, C.; Wu, J.; Fleischmann, D.; Harvey, H.; Folio, L.R.; Summers, R.M.; Rubin, D.L.; Lungren, M.P. Preparing medical imaging data for machine learning. *Radiology* **2020**, *295*, 4–15. [CrossRef]
12. Richens, J.G.; Lee, C.M.; Johri, S. Improving the accuracy of medical diagnosis with causal machine learning. *Nat. Commun.* **2020**, *11*, 1–9. [CrossRef]
13. Zemblys, R.; Niehorster, D.C.; Komogortsev, O.; Holmqvist, K. Using machine learning to detect events in eye-tracking data. *Behav. Res. Methods* **2018**, *50*, 160–181. [CrossRef] [PubMed]
14. Grewal, P.S.; Oloumi, F.; Rubin, U.; Tennant, M.T. Deep learning in ophthalmology: A review. *Can. J. Ophthalmol.* **2018**, *53*, 309–313. [CrossRef] [PubMed]
15. Armstrong, G.W.; Lorch, A.C. A (eye): A review of current applications of artificial intelligence and machine learning in ophthalmology. *Int. Ophthalmol. Clin.* **2020**, *60*, 57–71. [CrossRef] [PubMed]
16. Wang, Z.; Keane, P.A.; Chiang, M.; Cheung, C.Y.; Wong, T.Y.; Ting, D.S.W. Artificial intelligence and deep learning in ophthalmology. *Artif. Intell. Med.* **2020**, 1–34. [CrossRef]
17. Liu, T.A.; Ting, D.S.; Paul, H.Y.; Wei, J.; Zhu, H.; Subramanian, P.S.; Li, T.; Hui, F.K.; Hager, G.D.; Miller, N.R. Deep learning and transfer learning for optic disc laterality detection: Implications for machine learning in neuro-ophthalmology. *J. Neuro-Ophthalmol.* **2020**, *40*, 178–184. [CrossRef]
18. Jais, F.N.; Che Azemin, M.Z.; Hilmi, M.R.; Mohd Tamrin, M.I.; Kamal, K.M. Postsurgery Classification of Best-Corrected Visual Acuity Changes Based on Pterygium Characteristics Using the Machine Learning Technique. *Sci. World J.* **2021**, *2021*, 6211006. [CrossRef]
19. Ryu, H.; Ryu, H.S.; Wallraven, C. Analysis of Vision Acuity (VA) using Artificial Intelligence (AI): Comparison of Machine Learning Models and Proposition of an Optimized Model. *J. Korea Soc. Vis. Sci.* **2020**, *22*, 229–236. [CrossRef]
20. Rohm, M.; Tresp, V.; Müller, M.; Kern, C.; Manakov, I.; Weiss, M.; Sim, D.A.; Priglinger, S.; Keane, P.A.; Kortuem, K. Predicting visual acuity by using machine learning in patients treated for neovascular age-related macular degeneration. *Ophthalmology* **2018**, *125*, 1028–1036. [CrossRef]

21. Alexeeff, S.E.; Uong, S.; Liu, L.; Shorstein, N.H.; Carolan, J.; Amsden, L.B.; Herrinton, L.J. Development and Validation of Machine Learning Models: Electronic Health Record Data To Predict Visual Acuity After Cataract Surgery. *Perm. J.* **2020**, *25*, 25. [CrossRef]
22. Li, H.; Chutatape, O. Fundus image features extraction. In Proceedings of the 22nd Annual International Conference of the IEEE Engineering in Medicine and Biology Society (Cat. No. 00CH37143), Chicago, IL, USA, 23–28 July 2000; Volume 4, pp. 3071–3073.
23. Mateen, M.; Wen, J.; Song, S.; Huang, Z. Fundus image classification using VGG-19 architecture with PCA and SVD. *Symmetry* **2019**, *11*, 1. [CrossRef]
24. Xu, K.; Feng, D.; Mi, H. Deep convolutional neural network-based early automated detection of diabetic retinopathy using fundus image. *Molecules* **2017**, *22*, 2054. [CrossRef] [PubMed]
25. Cheng, X.; Feng, X.; Li, W. Research on Feature Extraction Method of Fundus Image Based on Deep Learning. In Proceedings of the 2020 IEEE 3rd International Conference on Automation, Electronics and Electrical Engineering (AUTEEE), Shenyang, China, 20–22 Novemebr 2020; pp. 443–447.
26. Trucco, E.; Ruggeri, A.; Karnowski, T.; Giancardo, L.; Chaum, E.; Hubschman, J.P.; Al-Diri, B.; Cheung, C.Y.; Wong, D.; Abramoff, M.; et al. Validating retinal fundus image analysis algorithms: Issues and a proposal. *Investig. Ophthalmol. Vis. Sci.* **2013**, *54*, 3546–3559. [CrossRef] [PubMed]
27. Ho, C.Y.; Pai, T.W.; Chang, H.T.; Chen, H.Y. An atomatic fundus image analysis system for clinical diagnosis of glaucoma. In Proceedings of the 2011 International Conference on Complex, Intelligent, and Software Intensive Systems, Seoul, Korea, 30 June–2 July 2011; pp. 559–564.
28. Simonyan, K.; Zisserman, A. Very Deep Convolutional Networks for Large-Scale Image Recognition. *arXiv* **2015**, arXiv:1409.1556.
29. Tan, M.; Le, Q.V. EfficientNet: Rethinking Model Scaling for Convolutional Neural Networks. *arXiv* **2019**, arXiv:1905.11946.
30. Thurnhofer-Hemsi, K.; López-Rubio, E.; Molina-Cabello, M.A.; Najarian, K. Radial basis function kernel optimization for Support Vector Machine classifiers. *arXiv* **2020**, arXiv:2007.08233.
31. Scholkopf, B. The kernel trick for distances. In *Advances in Neural Information Processing Systems*; MIT Press: Cambridge, MA, USA, 2001; pp. 301–307.

Article

A Novel Bayesian Linear Regression Model for the Analysis of Neuroimaging Data

Albert Belenguer-Llorens [1], Carlos Sevilla-Salcedo [1], Manuel Desco [2,3,4,5], Maria Luisa Soto-Montenegro [3,4,*] and Vanessa Gómez-Verdejo [1,*]

1   Department of Signal Processing and Communications, University Carlos III of Madrid Leganés, 28911 Leganés , Spain; abelenguer@tsc.uc3m.es (A.B.-L.); casevill@pa.uc3m.es (C.S.-S.)
2   Department of Bioengineering and Aerospace Engineering, University Carlos III of Madrid Leganés, 28911 Leganés, Spain; desco@hggm.es
3   CIBER of Mental Health (CIBERSAM), 28029 Madrid, Spain
4   Instituto de Investigación Sanitaria Gregorio Marañón, 28007 Madrid, Spain
5   Centro Nacional de Investigaciones Cardiovasculares (CNIC), 28029 Madrid, Spain
*   Correspondence: marisa@hggm.es (M.L.S.-M.); vanessag@ing.uc3m.es (V.G.-V.)

Abstract: In this paper, we propose a novel Machine Learning Model based on Bayesian Linear Regression intended to deal with the low sample-to-variable ratio typically found in neuroimaging studies and focusing on mental disorders. The proposed model combines feature selection capabilities with a formulation in the dual space which, in turn, enables efficient work with neuroimaging data. Thus, we have tested the proposed algorithm with real MRI data from an animal model of schizophrenia. The results show that our proposal efficiently predicts the diagnosis and, at the same time, detects regions which clearly match brain areas well-known to be related to schizophrenia.

Keywords: Bayesian learning; neuroimaging; feature selection; kernel formulation; mental disorders; schizophrenia; MRI

Citation: Belenguer-Llorens, A.; Sevilla-Salcedo, C.; Desco, M.; Soto-Montenegro, M.L.; Gomez-Verdejo, V. A Novel Bayesian Linear Regression Model for the Analysis of Neuroimaging Data. *Appl. Sci.* **2022**, *12*, 2571. https://doi.org/10.3390/app12052571

Academic Editors: Alexander E. Hramov, Kyungtae Kang, Hyo-Joong Suh and Junggab Son

Received: 7 December 2021
Accepted: 21 February 2022
Published: 1 March 2022

Publisher's Note: MDPI stays neutral with regard to jurisdictional claims in published maps and institutional affiliations.

1. Introduction

Neuroimaging has undergone a major breakthrough in recent years and has helped in the diagnosis, prognosis, and treatment monitoring of psychiatric disorders. The clinical diagnosis of these disorders is troublesome due to the lack of specific biomarkers [1] and to the fact that many of them share clinical features, thus hindering an accurate diagnosis. Specifically, schizophrenia is one of the most complex pathologies to diagnose [2] since it is commonly confused with other psychotic disorders in up to 20% of cases [3]. As consequence, new tools for the diagnosis of mental disorders are emerging [4,5].

Machine Learning (ML) techniques have emerged as a promising tool for the analysis of neuroimaging data. These algorithms are capable of analyzing any data source, either images (structural or functional), genetic information [6] or behavioral information [7], to carry out an automatic diagnosis of the pathology. Recent approaches based on Support Vector Machine algorithm (SVM) have been applied in Magnetic Resonance Imaging (MRI), showing great results in this field and detecting relevant brain areas involved in the pathology, as well as inferring new useful biomarkers for their diagnosis [8–10]. However, although these models have provided accurate results for automatic classification, the lack of interpretability in their results prevents the characterization of the pathology. In particular, in contexts where only a few features are relevant for the problem, it is advisable to detect the informative variables and eliminate the useless ones. For this reason, many authors combine ML models with Feature Selection (FS) approaches, such as the Recursive Feature Elimination (RFE) [11], consisting of the direct elimination of the less representative features, methods based on decision tree formulations, such as Random Forest Importance (RFI) [12,13], or embedded approaches which include L1 or

L1–L2 regularizations to promote sparsity, such as Lasso and elastic-net algorithms [14,15]. Nevertheless, in neuroimaging, we have to deal with large datasets, where the number of cases is significantly smaller than the number of variables, and many of these approaches fail in this scenario, tending to over-fit. To avoid this problem, some authors propose Bayesian approaches but work over a reduced set of features [16–18], whereas others point to the use of more refined techniques better adapted to the problem needs [19–21].

To overcome these limitations, we present a novel formulation for the Bayesian Linear Regression model. Our proposal, called the Dual Bayesian Linear regression model with Feature Selection (DBL-FS), is formulated to work efficiently with a reduced number of samples characterized in high-dimensional spaces, e.g., neuroimaging data. For this purpose, the model is formulated in the dual space and simultaneously includes an Automatic Relevance Determination (ARD) prior over the primal weights to provide the model with FS capabilities so that it can remove irrelevant input features. Here, we have tested our formulation on rodent data in an animal model of schizophrenia that show similar brain anatomical deficits than patients with schizophrenia [22–24]. One advantage of using rodent data is a more solid knowledge of the ground truth due to the controlled experimental induction of the pathology.

2. Materials

Rodent MRI data were obtained from the Biomedical Imaging and Instrumentation Group (Biig) of the Gregorio Marañón Hospital. The dataset consisted of 53 rat brain MRI images divided into two groups: healthy rats (N = 24) and pathological rats (N = 29). Pathology was induced by the administration of the viral mimic polyriboinosinic-polyribocytidilic acid (poly I:C) in gestational day 15 to pregnant Wistar rats, since maternal immune stimulation (MIS) is associated with increased risk of onset of schizophrenia in the offspring, with behavioral abnormalities as well as neurophysiological and morphological traits. Model details can be found elsewhere [25–27].

All images were preprocessed following the standard preprocessing pipeline in neuroimaging research, using the processing toolbox of the Statistical Parametric mapping software (SPM12) [28], as shown in Figure 1. Output consisted of: White Matter (WM), Gray Matter (GM), and CerebroSpinal Fluid (CSF) regions, with 464,487, 582,467, and 30,702 voxels, respectively.

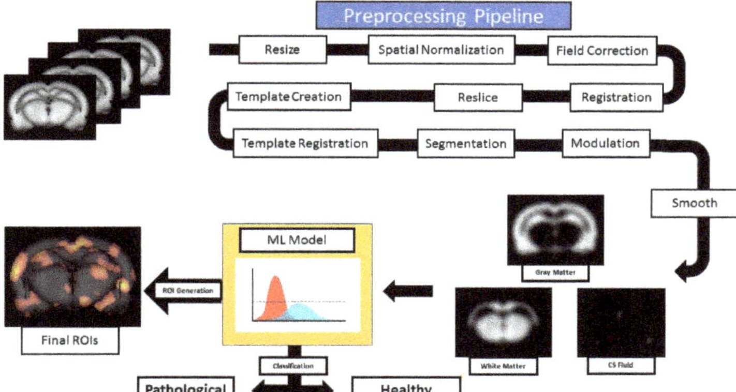

Figure 1. MRI pipeline for data processing [28]. First, images were corrected for field homogeneity, resized by a factor of 10 and spatially normalized to create a custom template based on a Wistar rat brain template [29]. All images were resliced to this custom template and were segmented into gray matter (GM), white matter (WM), and cerebrospinal fluid (CSF). Later, all images were modulated using the Jacobian determinants and smoothed with a 10-mm FWHM Gaussian kernel. Finally, the segmented tissues were processed by the ML model to classify them into healthy and pathological subjects and to identify the brain areas relevant to this decision.

3. Methods

This section introduces the formulation of the proposed Dual Bayesian Linear regression model with Feature Selection (DBL-FS). Later, we also introduce some reference methods that we will use as baselines to show the advantages of the proposed approach together to the experimental setup.

3.1. A Dual Bayesian Linear Regression Model with Feature Selection

3.1.1. Model Definition

The proposed model borrows some ideas from the Bayesian Principal Component Analysis (BPCA) [30] and Bayesian Canonical Correlation Analysis (BCCA) [31] algorithms to endow a Bayesian Linear Regression (BLR) [32] with a dual formulation able to carry out automatic feature selection over the primal variables. This relies on including an ARD prior over the weight matrices to automatically infer the feature relevance in the input feature space by assigning higher/lower relevance values when there are more/less relevant features. Meanwhile, the model works with a formulation in the dual space. In turn, this allows the model to efficiently deal with large data problems by working in the data space while it applies a feature selection over the variable space. In addition, we can exploit the DBL-FS Bayesian formulation to facilitate including prior expert knowledge to guide the FS process. This way, we can guide the learning process and compensate the limited number of samples to train the model.

To define the model, let us consider $\mathbf{X}$ as the observation matrix with the MRI information of N subjects; this way, each row, $\mathbf{x}_{n,:}$ for $n = 1, \ldots, N$, is a D-dimensional vector containing the brain image of the n-th subject, and each column, $\mathbf{x}_{:,d}$ for $d = 1, \ldots, D$, contains the information of the d-th voxel over the N subjects. On the other hand, the column vector $\mathbf{y}$ represents the diagnosis labels (control or schizophrenic) for the N subjects under study. Although each label, y_n, belongs to the set $\{0, 1\}$ (indicating the subject is control or not), for the model formulation, we consider $y_n \in \Re$, and thus, we will generalize the model for regression problems. Later, we will apply a threshold over the model output to classify each subject into one of two categories.

3.1.2. Generative Model

As the graphical model of Figure 2 shows, the generative model of DBL-FS considers that each datum, $\mathbf{x}_{n,:}$, is combined with a weight vector $\mathbf{w}$ plus some Gaussian noise to generate the output variable:

$$y_n = \mathbf{x}_{n,:}\mathbf{w} + \eta, \tag{1}$$

where η is a Gaussian noise with zero mean and precision τ. In turn, the noise precision is modeled with a gamma distribution with parameters a_0^τ, b_0^τ:

$$\tau \sim \Gamma(a_0^\tau, b_0^\tau) \tag{2}$$

In addition, DBL-FS considers that the weight associated to the d-th input feature follows a normal distribution:

$$w_d \sim \mathcal{N}\left(0, \alpha_d^{-1}\right) \quad d = 1, \ldots, D \tag{3}$$

where its precision, α_d, is modeled with a gamma distribution as:

$$\alpha_d \sim \Gamma(a_0^\alpha, b_0^\alpha) \quad d = 1, \ldots, D \tag{4}$$

This ARD prior over w_d allows us to obtain the relevance over the elements of $\mathbf{w}$, and therefore, DBL-FS is capable of automatically setting to zero the features that are irrelevant for the problem.

As the model will have to work with MRI data, composed by few samples (less than 100) and tens or hundreds of thousands of voxels, it is clear that working in the primal space is not the most efficient way to proceed. So, we propose to reformulate the model making use of the Representer Theorem [33] (RT). That is, as the RT states that the primal weights of any regression model resulting from minimizing an empirical error (risk) can be expressed as a linear combination of the input data and its equivalent dual coefficients, we can express $\mathbf{w}$ as:

$$\mathbf{w} = \mathbf{X}^T \mathbf{a} \tag{5}$$

where $\mathbf{a}$ is a vector of length N containing the dual variables. As we will see later (see Equation (19)), the lower bound that maximizes our variational inference is equivalent to minimizing an empirical cost. This way, the model can be formulated to work in the dual space as:

$$y_n = \mathbf{k}_{n,:}\mathbf{a} + \eta, \tag{6}$$

where $\mathbf{k}_{n,:}$ denotes to the n-th row of the linear kernel matrix $\mathbf{K} = \mathbf{X}\mathbf{X}^T$. This way, with the dual formulation, the target variables y_n, for $n = 1, \ldots, N$, are modeled as:

$$y_n \sim \mathcal{N}\left(\mathbf{k}_{n,:}\mathbf{a}, \tau^{-1}\right) \quad n = 1, \ldots, N. \tag{7}$$

With this new formulation, the model will be able to work in the space of $\mathbf{a}$, where only N parameters have to be inferred. Thus, model complexity and overfitting risks are drastically reduced, as long as we are able to maintain the feature relevance determination over $\mathbf{w}$, providing the model with feature selection capabilities.

Finally, it is important to note that the model formulation does not need to specifically include the distribution of $\mathbf{a}$ since the relation between $\mathbf{w}$ and $\mathbf{a}$ is deterministic, and therefore, the statistical characterization of $\mathbf{w}$ is also characterizing $\mathbf{a}$.

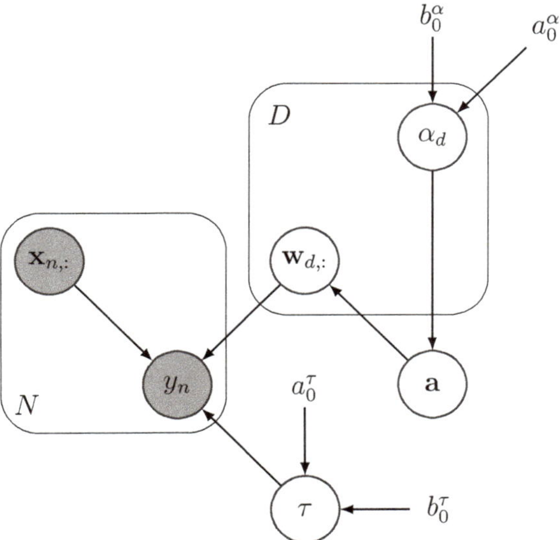

Figure 2. Plate diagram for the DBL-FS graphic model. Grey circles denote observed variables, white circles unobserved variables. Model hyperparameters do not have a circle.

3.1.3. Variational Inference

Once the generative model is defined, we should evaluate the posterior distribution of the variables to estimate their optimum values. Although, in this case, the posterior distribution is not tractable, we can use variational inference together with the mean-field technique [34] to find an approximation to this posterior $q(\boldsymbol{\Theta}) \approx p(\boldsymbol{\Theta}|\mathbf{y}, \mathbf{X})$, where $\boldsymbol{\Theta}$ contains all model variables. Then, we can define a Lower Bound (LB) using the Kullback–

Leibler divergence between the posterior and its approximation; so, maximizing this LB, we can obtain the optimum model parameters. Therefore, using the mean-field approximation to factorize over the posterior, we obtain:

$$p(\boldsymbol{\Theta}|\boldsymbol{y}, \mathbf{X}) \approx q(\boldsymbol{\Theta}) = q(\mathbf{w})q(\mathbf{a})q(\boldsymbol{\alpha})q(\tau), \tag{8}$$

and we can determine each approximated distribution by calculating:

$$\ln(q_j^*) = \mathbb{E}_{-q_j}[\ln(p(\mathbf{X}, \boldsymbol{y}, \boldsymbol{\Theta}))] + const, \tag{9}$$

where $\mathbb{E}_{-q_j}$ implies that we calculate the expectation over all random variables except the j-th variable, and $p(\mathbf{X}, \boldsymbol{y}, \boldsymbol{\Theta})$ is the joint probability.

Therefore, we can apply (9) to the joint probability for each random variable to obtain the model update rules. Firstly, the distribution of the dual weights $\mathbf{a}$ is:

$$q(\mathbf{a}) = \mathcal{N}(\mathbf{a}|\langle \mathbf{a}\rangle, \boldsymbol{\Sigma}_a), \tag{10}$$

with mean and variance determined by:

$$\langle \mathbf{a}\rangle = \langle \tau\rangle \boldsymbol{\Sigma}_a \mathbf{K}^T y \tag{11}$$

$$\boldsymbol{\Sigma}_a^{-1} = \mathbf{X}\mathrm{diag}(\langle \boldsymbol{\alpha}\rangle)\mathbf{X}^T + \langle \tau\rangle \mathbf{K}^T \mathbf{K}, \tag{12}$$

where $\mathrm{diag}(\langle \boldsymbol{\alpha}\rangle)$ represents an identity matrix with vector α as the diagonal. The distribution of variable $\boldsymbol{\alpha}$ is:

$$q(\boldsymbol{\alpha}) = \Gamma(\boldsymbol{\alpha}|a_\alpha, \mathbf{b}_\alpha), \tag{13}$$

with parameters

$$a_\alpha = a_0^\alpha + \frac{D}{2} \tag{14}$$

$$\mathbf{b}_\alpha = b_0^\alpha + \frac{1}{2}\mathrm{diag}(\mathbf{X}^T\langle \mathbf{aa}^T\rangle \mathbf{X}), \tag{15}$$

where α_0^α and β_0^α are hyperparameters, and the operator diagonal returns a column vector formed by the main diagonal of the matrix. Moreover, the distribution of the noise precision τ is given by:

$$q(\tau) = \Gamma(\tau|a_\tau, b_\tau), \tag{16}$$

with parameters

$$a_\tau = a_0^\tau + \frac{N}{2} \tag{17}$$

$$b_\tau = b_0^\tau + \frac{1}{2}\left(\sum_{n=1}^{N} y_n^2 - 2Tr\{\boldsymbol{y}^T\mathbf{K}\langle \mathbf{a}\rangle\} + Tr\{\mathbf{K}^T\mathbf{K}\langle \mathbf{aa}^T\rangle\}\right), \tag{18}$$

where α_0^τ and β_0^τ are hyperparameters, and $Tr\{\}$ is the trace operator. See Appendix A for the full development of these mean field distribution approximations.

Once we have defined the different distributions, the model updates the different random variables in an iterative coordinate-ascent-like optimization where the distribution of each factor is obtained using (10) to (22). This optimization process is guided by the *LB* cost function defined as:

$$LB = const + \sum_{n=1}^{N}\left(\frac{D}{2} + a_0^\alpha + 1\right)\ln(b_\alpha) - \left(\frac{D}{2} + a_0^\tau + 1\right)\ln(b_\tau) - \frac{D}{2}\ln(|\boldsymbol{\Sigma}_a|), \tag{19}$$

where we analyze its convergence to stop the distribution parameters update. See Appendix B for the full development of the LB.

For an efficient optimization of the model, in practice, we will work in the dual space updating the Equations (10) to (18). However, when the model convergence is reached, we can obtain the approximate posterior distribution of $\mathbf{w}$ as:

$$q(\mathbf{w}) = \mathcal{N}(\mathbf{w}|\langle\mathbf{w}\rangle, \boldsymbol{\Sigma}_w),$$ (20)

with parameters

$$\langle\mathbf{w}\rangle = \langle\mathbf{a}\rangle\mathbf{X}$$ (21)

$$\boldsymbol{\Sigma}_w = \mathbf{X}^T\boldsymbol{\Sigma}_a\mathbf{X}.$$ (22)

Once the model is trained, we can analyze the distribution of $\mathbf{w}$ and check which feature components are zero and, therefore, are eliminated, having an automatic selection of the relevant input voxels. This is due to the fact that, despite working in the dual space, the precision of $\mathbf{w}$ components, α, is considered in the distribution of $\mathbf{a}$ (see Equation (12)).

Moreover, the inclusion of a prior over α (see Equation (13)) in the generative model has an additional advantage since we can use it to adapt the prior distribution of $\mathbf{w}$ and include expert knowledge in the model. Thus, in case we want to add more relevance to a particular region (for instance, a neurobiologically meaningful Region of Interest (ROI)), we can initialize the parameters b_α associated to the voxels of this region with higher values than the rest to promote that the distribution of $\mathbf{w}$ has also higher values for these voxels. Otherwise, if we do not want to include this expert knowledge, this variable will be uniformly initialized over all voxels.

3.2. Baselines

Here, we present the baseline methods used during the experimental section, whose performances will be compared with those of the proposed DBL-FS model. In particular, we considered three approaches, one baseline aimed to solve regression problems (as DBL-FS) and the other two methods specifically designed to solve classification tasks:

- As the first baseline, we included a regression Gaussian Process (GP) [35], using the implementation provided by the GPy library (available at github accessed on 9 December 2021). We have selected this model since it allows us to define lineal kernels with ARD, so that we can work in the dual space and learn the relevance of the different input features.
- Next, we included an SVM [36] with a linear kernel using the scikit-learn library [37] to also optimize the model in the dual space.
- The last selected baseline is the recently proposed adaptation of Sparse Semi-supervised Heterogeneous Interbattery Bayesian Analysis (SSHIBA) [38] to work in the dual space, the Kernelized SSHIBA (KSSHIBA) [39] is available at github accessed on 29 March 2021. This algorithm can simultaneously combine different data sources or views (in our case, different tissues) in a common latent space providing a low-dimensional representation of the data. In addition, this model can also include an additional output view to categorically model the target variable (patient or control sample), as well as a linear kernel with ARD coefficients over the input features (equivalent to the GP configuration).

Both GP and KSSHIBA use an ARD to determine the relevance of the input features, but they do not have a prior distribution or constraint to force their input weights to be sparse and, therefore, obtain a real FS. Meanwhile, DBL-FS imposes sparsity with the Gamma prior to actually promote zero values in the model weights which, in turn, automatically eliminates the least relevant features.

Furthermore, it is important to mention that deep learning models are not included in this study, as these methods are severely limited by the sample size required to learn the model parameters. Therefore, although models such as convolutional neural networks have promising results in image analysis, they also pose serious challenges when working with datasets of small sample size. Furthermore, we have explored other baselines such as random forests but have not included the results due to their poor results. Nevertheless, all the methods under study will be evaluated with different configurations to be able to analyze different properties and, hence, to carry out an extensive study and analyze their advantages and disadvantages in comparison to our model.

3.3. Experimental Setup

MRI data were standardized to zero mean and unitary standard deviation. As we have a reduced number of subjects (only 53 samples), we have used a Leave-One-Out (LOO) framework to evaluate the model performance. This way, we have trained as many models as available samples, using in each training partition all the subjects except one, which was used afterwards for testing. Then, to evaluate the model performance, we present the results in terms of average accuracy, that is, the percentage of correctly classified test subjects computed over all LOO iterations. Furthermore, since the performance of some methods depends on their initialization we repeated the LOO process 10 times (with different initializations) and depicted the average accuracy over them in order to obtain more statically significant results.

To complete the performance analysis, the result table includes the final number of voxels selected by each model (and their percentage with respect to the total), computed as the average number of voxels used by each model for each LOO iteration and each run.

Regarding the different models under study, we considered several configurations to carry out a more comprehensive analysis and more adequate evaluation of the different methods.

For GPs, we have considered two versions: (1) the standard GP with a linear kernel, denoted as GP, and (2) the previous GP but including ARD capabilities and an FS stage. That is, we first trained a GP with ARD and analyzed the ARD coefficients to select the most relevant features, and then trained a standard GP with the chosen features. Thus, this two-step approach provided a GP with FS capabilities, denoted as GP+FS. For this pruning, we selected the 25% most relevant features in order to compare the performance of this method with DBL-FS. In addition, as both DBL-FS and GPs were formulated for regression problems and our predictive task is a binary classification (0 or 1), we set the threshold to 0.5.

We have implemented two different approaches for SVMs: (1) a standard SVM with linear kernel and (2) an SVM with a Multi-Kernel Learning (MKL) strategy, denoted as MKL-SVM. In the latter case, we independently considered the different tissues (GM, CSF, and WM) and a different linear kernels for each of them, and subsequently, the model learned the combination of these three kernels, including two parameters for their combination. These parameters were defined as scalars multiplying each kernel term, and a subsequent inner LOO was used to find their optimal values. Thus, the defined combinations coefficients gave more or less relevance to each kernel (therefore, to each tissue), providing additional flexibility to the model.

For KSSHIBA, we have included two versions, similarly to what was done in GP: (1) the standard KSSHIBA model and (2) a two-stage version of KSSHIBA (denoted as KSSHIBA+FS), in which KSSHIBA was first trained with ARD functionality, and subsequently, we selected the most relevant features to train the model using this subset of features. For these experiments, we initially had 1000 latent factors, from which the model will automatically prune the irrelevant ones. For FS, we kept the highest 25% of voxels equivalently to the number of selected features from the DBL-FS model.

Finally, we have also defined two approaches for the DBL-FS model, with and without expert knowledge. In the latter one, we have equally initialized the ARD prior for all voxels, setting the parameters a and b of random variable α to 2 and 1, respectively. In the expert knowledge case (denoted as DBL-FS+EK), we have initialized the parameters a and b in such a way that the areas of the prefrontal cortex, ventral hippocampus, and lateral ventricles (which are known to be more intensely affected [23]) had more relevance than the rest. In particular, parameter a was set to 50 and parameter b was fixed to either 1 or 0.001, depending on whether the voxel belonged to the indicated ROIs or not.

4. Experimental Results

Table 1 shows the LOO accuracy results for the classification problem together with the number of selected voxels (the approaches without FS used 100%). Despite using different initializations in the evaluated models, the results were stable across them with a negligible standard deviation, showing that the initialization hardly influences the results. For this reason, we did not include the standard deviation in Table 1. The results show that GPs, KSSHIBA+FS, and MKL-SVM obtained the worst classification accuracies, while SVM and KSSHIBA achieved the best performances among the baselines. However, DBL-FS and DBL-FS+EK still obtained an improvement of 5.7% in terms of accuracy over the best baseline while learning the most restrictive selection mask. From these results, we need to highlight that (1) KSSHIBA obtained a predictive performance similar to that of SVM while summarizing the information of the original data (distributed in more than 10^6 voxels) in only nine latent variables, and (2) MKL-SVM showed worse results than the standard SVM, probably due to the higher number of hyperparameters it needed to learn in order to perform the MKL, which may be causing overfitting.

Table 1. Performance of the different methods under study showing the model accuracy and the number of selected voxels (with their percentages with respect to the total). In addition, models with the best accuracy have been highlighted in bold and placed at the bottom of the table, which corresponds to the proposed DBL-FS approaches.

Experiment	Accuracy	# Selected Voxels
GP	67.9%	1,077,656 (100%)
GP+FS	67.9%	269,414 (25%)
SVM	71.6%	1,077,656 (100%)
MKL-SVM	67.9%	1,077,656 (100%)
KSSHIBA	69.8%	1,077,656 (100%)
KSSHIBA+FS	64.1%	269,414 (25%)
DBL-FS	**77.3%**	**287,996 (26.72%)**
DBL-FS+EK	**77.3%**	**242,754 (22.52%)**

Figure 3 shows the brain areas selected by the methods with FS capabilities. As each voxel has an associated weight, the image masks represent the absolute value of these weight magnitudes for the selected of voxels as an indicator of the voxel relevance. In addition, since we have a model for each LOO iteration, Figure 3 displays the average values of these relevances (over all LOO iterations) and includes a normalization of their scales to the range $(0, 1)$ to simplify their analysis. As a result, we can observe that the GP-FS selected meaningless voxels in neurobiological terms while KSSHIBA detected well-defined areas corresponding to only WM tissues. Finally, the DBL-FS and DBL-FS+EK approaches obtained well-defined regions in the GM and WM tissues and the CSF, which are interpretable in neurobiological terms. Although both methods provided similar selections, DBL-FS+EK selected a reduced set of voxels.

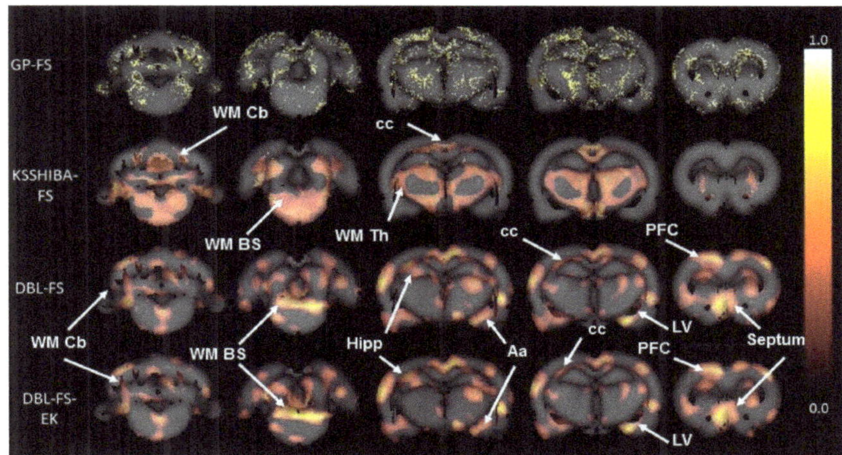

Figure 3. Brain masks obtained by the FS of each model. Colors are defined in a linear scale and associated to the relevance of the voxel (white: more relevant; dark red: less relevant). GP-FS model yields meaningless results in neurobiological terms, where no anatomical regions can be identified. KSSHIBA-FS model only identifies brain areas related to WM deficits in schizophrenia. Both DBL-FS and DBL-FS+EK learn similar relevance in WM, GM, and CSF brain areas of interest in schizophrenia, such as the hippocampus (hipp), prefrontal cortex (PFC), amygdala (Aa), septum, lateral ventricles (LV), corpus callosum (cc), WM cerebellar (WM Cb), and WM brainstem (WM BS) fibers.

5. Discussion

This study shows, for the first time, the great advantage of using DBL-FS for the detection and characterization of the morphometric brain changes in a rodent model of schizophrenia. This Bayesian model was adapted for neuroimaging data, characterized by a low sample-to-variable ratio (53 samples vs. 1,077,656 voxels in our case) relying on a dual formulation of the Bayesian Linear Regression model. Furthermore, as the main novelty of this proposal, we combine this dual formulation with a prior over the primal weights to learn the feature relevance over the input features, forcing an automatic FS. Finally, we can exploit the Bayesian nature of the model to include specific prior knowledge to guide the learning process and counterweight the limitations caused by the low sample size of the problem.

Thus, the comparison in terms of performance with the baselines provides clear evidence of the promising results of the proposed DBL-FS model in the characterization of neuroimaging data in mental disorders. Note that DBL-FS is able to largely outperform the baselines in prediction accuracy, showing an advantage of 5.7% in terms of accuracy over the best baseline. In addition, DBL-FS is the only method capable of detecting regions within the three brain tissues that are known to be relevant in the biology of schizophrenia. In this sense, the relevance learned by the GPs is inconsistent between the different LOO iterations, generating a scattered voxel selection and, therefore, a non-localized, unreliable, and uninterpretable mask. On the other hand, KSSHIBA provides a consistent result but only finds relevant regions within WM tissue and, therefore, ignores relevant regions and includes some irrelevant areas.

Analyzing in detail the regions selected by the DBL-FS and DBL-FS+Ek models, we can verify that these areas belong to brain regions whose morphometric changes have been related to schizophrenia, based on the literature. First, as for CSF, the areas with the greatest weight were the most frontal areas of the lateral ventricles and the third ventricle. One of the morphometric hallmarks in schizophrenia is the enlargement of the ventricles [23,40], which is consistent with the learned selection. Second, regarding GM, our model clearly defined anatomical areas, such as the prefrontal cortex (PFC), hippocampus, amygdala, and septum, some of them in both hemispheres. Numerous studies have demonstrated the relevance of the morphological changes of these areas in mental disorders [41,42] together with the disconnection and lack of symmetry between both cerebral hemispheres [43,44]. Similar volumetric abnormalities have also been reported for the animal model used in this study [24,45]. In addition, the method also detected the medial septum, which plays a significant role in dopamine-related disorders such as schizophrenia [46,47] and addictions [48–50], which highlights the relevance of this structure in mental disorders. Regarding WM, the method found three well-defined brain areas, the frontal part of corpus callosum and WM tracts of the brainstem and the cerebellum [51,52].

Regarding the inclusion of expert knowledge by means of the α prior, it reveals two interesting behaviors. First, it demonstrates the robustness and potential of the standard DBL-FS since it is able to obtain similar accuracy and roughly similar brain masks to its DBL-FS+EK extension without the need for expert information. Second, the possibility of including expert knowledge makes the model converge faster, and it also refines the brain region selection. It is important to note that, although the expert knowledge guides the inference process, the model is also learning from the data, allowing it to redefine the initial expert knowledge into a specific set of voxels. For instance, looking at Figure 3, we can see that using expert knowledge, we obtain a higher relevance associated with the core of the WM brainstem and hippocampal areas.

6. Conclusions

This article shows a novel Bayesian approach using linear regression to characterize neuroimaging data, tested in an animal model of schizophrenia. The proposed DBL-FS+EK model allowed us to efficiently work with neuroimaging data, characterized by a low sample-to-variable ratio. This is achieved by taking advantage of its Bayesian formulation to work in the dual space while learning a voxel importance for feature selection. Furthermore, the use of a specific prior to force sparsity can be combined with expert knowledge to guide the model. The proposed model was analyzed using MRI data from a rodent model of a schizophrenia database and compared to different baselines. The results provided an outstanding classification performance of DBL-FS+EK, improving the accuracy of the second best classifier, SVM, in ~6%. Furthermore, looking at the selected voxels and their associated relevance, we can confirm that the proposed model is able to detect biologically relevant areas for the characterization of this disease, as it clearly agrees with known literature.

Author Contributions: Conceptualization, V.G.-V. and M.L.S.-M.; methodology, V.G.-V. and C.S.-S.; software, A.B.-L. and C.S.-S.; validation, M.L.S.-M. and A.B.-L.; formal analysis, A.B.-L., V.G.-V. and C.S.-S.; investigation, V.G.-V., M.L.S.-M. and C.S.-S.; resources, M.L.S.-M.; data curation, V.G.-V. and M.L.S.-M.; writing—review and editing, all; visualization, all; supervision, V.G.-V., M.L.S.-M. and C.S.-S.; project administration, V.G.-V. and M.L.S.-M.; funding acquisition, V.G.-V., M.L.S.-M. and M.D. All authors have read and agreed to the published version of the manuscript.

Funding: This paper is part of the project PID2020-115363RB-I00 funded by MCIN/AEI/10.13039/501100011033. A.B.-L.'s work is funded by the Community of Madrid through the "Excellence of University Teaching Staff" line of the Multi-year Agreement with UC3M (EPUC3M27), within the framework of the V PRICIT. M.L.S.-M.'s was supported by Ministerio de Ciencia, Innovación y Universidades, Instituto de Salud Carlos III (project number PI17/01766, and grant number BA21/00030), co-financed by European Regional Development Fund (ERDF), "A way to make Europe", CIBER de Salud Mental (project number CB07/09/0031), Delegación del Gobierno para el Plan Nacional sobre Drogas (project number 2017/085); Fundación Mapfre and Fundación Alicia Koplowitz. M.D.'s work was supported by Ministerio de Ciencia e Innovación (MCIN) and Instituto de Salud Carlos III (ISCIII) (PT20/00044). The CNIC is supported by the ISCIII, the MCIN and the Pro CNIC Foundation, and is a Severo Ochoa Center of Excellence (SEV-2015-0505).

Institutional Review Board Statement: Not applicable.

Informed Consent Statement: Not applicable.

Data Availability Statement: The dataset generated and analyzed during the current study are available from the corresponding author on reasonable request.

Conflicts of Interest: The authors declare no conflict of interest.

Appendix A. DBL-FS Variational Inference

This section explains in detail the development of the variational inference of the proposed DBL-FS indicated in the Methods section. In particular, here we present the calculation of the mean field approximation of the model parameters:

$$q(\mathbf{\Theta}) = q(\mathbf{w})q(\mathbf{a})q(\mathbf{\alpha})q(\tau), \tag{A1}$$

where each term is calculated applying Equation (9) to the joint probability for each random variable to obtain the updated model rules.

*Appendix A.1. Mean Field Approximation of **a***

Using the mean field approximation over variable $\mathbf{a}$, we find that the logarithm of its approximate posterior is:

$$\ln\left(q(\mathbf{a})\right) = \mathbb{E}[\ln\left(p(\boldsymbol{y}|\mathbf{X}, \mathbf{a}, \tau)\right)] + \mathbb{E}[\ln\left(p(\mathbf{w}|\boldsymbol{\alpha}, \mathbf{a})\right)] + const. \tag{A2}$$

If we develop the first term in the equation, we have:

$$\ln\left(p(\boldsymbol{y}|\mathbf{X}, \mathbf{a}, \tau)\right) = \sum_{n=1}^{N} \ln p(y_n|\mathbf{x}_{n,:}, \mathbf{a}, \tau) = \sum_{n=1}^{N} \ln\mathcal{N}\left(\mathbf{x}_{n,:}\mathbf{X}^T\mathbf{a}, \tau^{-1}\right)$$

$$= \sum_{n=1}^{N} \left(\frac{1}{2}\ln\left(\tau\right) - \frac{\tau}{2}(y_n - \mathbf{a}^T\mathbf{X}\mathbf{x}_{n,:}^T)(y_n - \mathbf{x}_{n,:}\mathbf{X}^T\mathbf{a})\right) + const$$

$$= \frac{N}{2}\ln\left(\tau\right) - \frac{\tau}{2}\sum_{n=1}^{N}\left(y_n^2 - 2y_n\mathbf{x}_{n,:}\mathbf{X}^T\mathbf{a} + \mathbf{a}^T\mathbf{X}\mathbf{x}_{n,:}^T\mathbf{x}_{n,:}\mathbf{X}^T\mathbf{a}\right) + const, \tag{A3}$$

and, calculating the expectation of this expression, we obtain:

$$\mathbb{E}_\tau[\ln\left(p(\boldsymbol{y}|\mathbf{X}, \mathbf{a}, \tau)\right)] = \frac{N}{2}\ln(\langle\tau\rangle) + \langle\tau\rangle\boldsymbol{y}\mathbf{K}\mathbf{a} - \frac{\langle\tau\rangle}{2}\mathbf{a}^T\mathbf{K}^T\mathbf{K}\mathbf{a} + const. \tag{A4}$$

Equivalently, the second term can be calculated as:

$$\ln\left(p(\mathbf{w}|\boldsymbol{\alpha}, \mathbf{a})\right) = \sum_{d=1}^{D}\ln p(w_d|\alpha_d, \mathbf{a}) = \sum_{d=1}^{D}\ln\mathcal{N}\left(0, \alpha_d^{-1}\right)$$

$$= \sum_{d=1}^{D}\left(\frac{1}{2}\ln\left(\alpha_d\right) - \frac{1}{2}\mathbf{a}^T\mathbf{x}_{:,d}\alpha_d\mathbf{x}_{:,d}^T\mathbf{a}\right) + const$$

$$= \frac{1}{2}\sum_{d=1}^{D}\ln\left(\alpha_d\right) - \frac{1}{2}\sum_{d=1}^{D}\left(\mathbf{a}^T\mathbf{x}_{:,d}\alpha_d\mathbf{x}_{:,d}^T\mathbf{a}\right) + const, \tag{A5}$$

and, we if we use the expectation, we have:

$$\mathbb{E}_\alpha[\ln\left(p(\mathbf{w}|\boldsymbol{\alpha}, \mathbf{a})\right)] = \frac{1}{2}\sum_{d=1}^{D}\ln(\langle\alpha_d\rangle) - \frac{1}{2}\mathbf{a}^T\mathbf{X}\mathrm{diag}(\langle\alpha\rangle)\mathbf{X}^T\mathbf{a} + const. \tag{A6}$$

Now, joining Equations (A4) and (A6), we obtain:

$$\ln\left(q(\mathbf{a})\right) = \langle\tau\rangle\boldsymbol{y}^T\mathbf{K}\mathbf{a} - \frac{\langle\tau\rangle}{2}\mathbf{a}^T\mathbf{K}^T\mathbf{K}\mathbf{a} - \frac{1}{2}\mathbf{a}^T\mathbf{X}\mathrm{diag}(\langle\alpha\rangle)\mathbf{X}^T\mathbf{a} + const. \tag{A7}$$

Therefore, we can identify the parameters of the q distribution on this equation, having:

$$q(\mathbf{a}) = \mathcal{N}(\mathbf{a}|\langle \mathbf{a}\rangle, \Sigma_a) \tag{A8}$$

where the mean is:

$$\langle \mathbf{a}\rangle = \tau \Sigma_a \mathbf{K}^{\mathsf{T}} \mathbf{y} \tag{A9}$$

and the variance is:

$$\Sigma_a^{-1} = \mathbf{X}\mathrm{diag}(\langle \boldsymbol{\alpha}\rangle)\mathbf{X}^{\mathsf{T}} + \langle\tau\rangle\mathbf{K}^{\mathsf{T}}\mathbf{K} \tag{A10}$$

Appendix A.2. Mean Field Approximation of $\boldsymbol{\alpha}$

Now, using the mean field approximation over variable $\boldsymbol{\alpha}$, we find that the logarithm of its approximate posterior is:

$$\ln\left(q(\boldsymbol{\alpha})\right) = \mathbb{E}[\ln\left(p(\mathbf{w}|\boldsymbol{\alpha}, \mathbf{a})\right)] + \mathbb{E}[\ln\left(p(\boldsymbol{\alpha})\right)] + const \tag{A11}$$

Developing the first term, we obtain:

$$\ln\left(p(\mathbf{w}|\boldsymbol{\alpha}, \mathbf{a})\right) = \frac{1}{2}\sum_{d=1}^{D}\ln\left(\alpha_d\right) - \frac{1}{2}\sum_{d=1}^{D}\mathrm{Tr}\{\mathbf{a}^{\mathsf{T}}\mathbf{x}_{:,d}\alpha_d\mathbf{x}_{:,d}^{\mathsf{T}}\mathbf{a}\} + const, \tag{A12}$$

and we can apply the expectation to obtain:

$$\mathbb{E}_{a,\tau}[p(\mathbf{w}|\boldsymbol{\alpha}, \mathbf{a})] = \frac{1}{2}\sum_{d=1}^{D}\ln\left(\alpha_d\right) - \frac{1}{2}\sum_{d=1}^{D}\alpha_d\mathrm{Tr}\{\mathbf{x}_{:,d}^{\mathsf{T}}\langle \mathbf{a}\mathbf{a}^{\mathsf{T}}\rangle\mathbf{x}_{:,d}\} \tag{A13}$$

If we look at the second term, we have

$$\ln\left(p(\boldsymbol{\alpha})\right) = \sum_{d=1}^{D}\left(-b_0^{\alpha}\alpha_d + (a_0^{\alpha} - 1)\ln\left(\alpha_d\right)\right) + const, \tag{A14}$$

where we can apply the expectation of the function as:

$$\mathbb{E}[\ln\left(p(\boldsymbol{\alpha})\right)] = \sum_{d=1}^{D}\ln(p(\alpha_d)) = \sum_{d=1}^{D}\left(-b_0^{\alpha}\alpha_d + (a_0^{\alpha} - 1)\ln\left(\alpha_d\right)\right) + const. \tag{A15}$$

Now, if we sum Equations (A13) and (A15), we obtain:

$$\ln(q(\boldsymbol{\alpha})) = \sum_{d=1}^{D}\left(\left(\frac{1}{2} + a_0^{\alpha} - 1\right)\ln(\alpha_d) - \frac{\alpha_d}{2}\left(\mathrm{Tr}\{\mathbf{x}_{:,d}^{\mathsf{T}}\langle \mathbf{a}\mathbf{a}^{\mathsf{T}}\rangle\mathbf{x}_{:,d}\} + 2b_0^{\alpha}\right)\right) + const. \tag{A16}$$

Thus, if we identify terms on the variable distribution, we have:

$$q(\alpha_d) = \prod_{d=1}^{D}\Gamma\left(\alpha_d|a_{\alpha_d}, b_{\alpha_d}\right), \tag{A17}$$

where the first parameter is:

$$a_{\alpha} = \frac{1}{2} + a_0^{\alpha}, \tag{A18}$$

and the second parameter can be expressed as:

$$b_{\alpha} = b_0^{\alpha} + \frac{1}{2}\mathrm{diag}(\mathbf{X}^{\mathsf{T}}\langle \mathbf{a}\mathbf{a}^{\mathsf{T}}\rangle\mathbf{X}). \tag{A19}$$

Appendix A.3. Mean Field Approximation of τ

Following the same steps as in the two previous approaches, we use the mean field approximation over variable τ to obtain the logarithm of the approximate posterior:

$$\ln\left(q(\tau)\right) = E[\ln\left(p(\boldsymbol{y}|\mathbf{X},\mathbf{a},\tau)\right)] + E[\ln\left(p(\tau)\right)] + const. \tag{A20}$$

Therefore, the first term on this equation is:

$$\ln\left(p(\boldsymbol{y}|\mathbf{X},\mathbf{a},\tau)\right) = \frac{N}{2}\ln(\tau) - \frac{\tau}{2}\left(\sum_{n=1}^{N} y_n^2 - 2\mathrm{Tr}\left\{\boldsymbol{y}^T\mathbf{K}\mathbf{a}\right\} + \mathrm{Tr}\left\{\mathbf{K}^T\mathbf{K}\mathbf{a}\mathbf{a}^T\right\}\right) + const, \tag{A21}$$

and, applying the expectation we obtain:

$$\mathbb{E}_a[\ln\left(p(\boldsymbol{y}|\mathbf{X},\mathbf{a},\tau)\right)] = \frac{N}{2}\ln(\tau)$$
$$- \frac{\tau}{2}\left(\sum_{n=1}^{N} y_n^2 - 2\mathrm{Tr}\left\{\boldsymbol{y}^T\mathbf{K}\langle\mathbf{a}\rangle\right\} + \mathrm{Tr}\left\{\mathbf{K}^T\mathbf{K}\langle\mathbf{a}\mathbf{a}^T\rangle\right\}\right) + const. \tag{A22}$$

The second term is defined as:

$$\mathbb{E}[\ln\left(p(\tau)\right)] = \ln\left(p(\tau)\right) = -b_0^\tau\tau + (a_0^\tau - 1)\ln\left(\tau\right) + const. \tag{A23}$$

Now, if we sum Equations (A22) and (A23), we obtain:

$$\ln\left(q(\alpha)\right) = \left(\frac{N}{2} + a_0^\tau - 1\right)\ln\left(\tau\right)$$
$$- \frac{\tau}{2}\left(\sum_{n=1}^{N} y_n^2 - 2\mathrm{Tr}\left\{\boldsymbol{y}^T\mathbf{K}\langle\mathbf{a}\rangle\right\} + \mathrm{Tr}\left\{\mathbf{K}^T\mathbf{K}\langle\mathbf{a}\mathbf{a}^T\rangle\right\} + 2b_0^\tau\right) + const. \tag{A24}$$

Therefore, following the same procedure as with the previous variables, we identify terms from the distribution and obtain:

$$q(\tau) = \Gamma(\tau|a_\tau, b_\tau), \tag{A25}$$

where the first parameter is:

$$a_\tau = \frac{N}{2} + a_0^\tau, \tag{A26}$$

and the second one is:

$$b_\tau = \frac{1}{2}\left(\sum_{n=1}^{N} y_n^2 - 2\mathit{Tr}\left\{\boldsymbol{y}^T\mathbf{K}\langle\mathbf{a}\rangle\right\} + \mathit{Tr}\left\{\mathbf{K}^T\mathbf{K}\langle\mathbf{a}\mathbf{a}^T\rangle\right\}\right) + b_0^\tau. \tag{A27}$$

Appendix B. Lower Bound Inference

As mentioned in the Methods section, we use the Kullback–Leibler divergence to first determine the similarities between two distribution where, for any two probability density functions $p(x)$ and $q(x)$, we have:

$$D_{KL} = \int q(x)\ln\frac{q(x)}{p(x)}dx \tag{A28}$$

In our case, if we particularize for the true posterior and the posterior approximation, the divergence can be expressed as:

$$D_{KL} = -\int q(\Theta) \ln\left(\frac{q(\Theta)}{p(\Theta|X)}\right) d\Theta = \int q(\Theta) \ln(q(\Theta)) d\Theta - \int q(\Theta) \ln(p(\Theta|X)) d\Theta$$
$$= \mathbb{E}_q[\ln(q(\Theta))] - \mathbb{E}_q[\ln(p(\Theta|X))]. \tag{A29}$$

Developing the conditional probability we obtain:

$$D_{KL} = \mathbb{E}_q[\ln(q(\Theta))] - \mathbb{E}_q[\ln(p(\Theta, X))] + \ln(p(X)). \tag{A30}$$

Due to the impossibility of working with this distribution because the marginal distribution $p(X)$ cannot be calculated, we use an Evidence Lower Bound (ELBO/LB) to this expression [34]. The LB is the divergence of negative KL plus $\ln(p(X))$; therefore, the greatest similarity between the two functions is achieved by maximizing this new measure. We can calculate the LB as:

$$L_q = -\int q(\Theta) \ln\left(\frac{q(\Theta)}{p(X, \Theta)}\right) d\Theta = \int q(\Theta) \ln(p(X, \Theta)) d\Theta - \int q(\Theta) \ln(q(\Theta)) d\Theta$$
$$= \mathbb{E}_q[\ln(p(X, \Theta))] - \mathbb{E}_q[\ln(q(\Theta))] \tag{A31}$$

In order to easily calculate this lower bound, we will separately calculate the terms related to $\mathbb{E}_q[\ln(p(X, \Theta))]$ and to the entropy in the following subsections.

Appendix B.1. Terms Associated to $\mathbb{E}_q[\ln(p(\mathbf{X}, \mathbf{y}, \Theta))]$

This first term of the lower bound would be composed by the following terms:

$$\mathbb{E}_q[\ln(p(\mathbf{X}, \mathbf{y}, \Theta))] = \mathbb{E}_q[\ln(p(\mathbf{X}))] + \mathbb{E}_q[\ln(p(\mathbf{w}\,|\,\boldsymbol{\alpha}, \mathbf{a}))] + \mathbb{E}_q[\ln(p(\boldsymbol{\alpha}))]$$
$$+ \mathbb{E}_q[\ln(p(\mathbf{y}\,|\,\mathbf{a}, \mathbf{X}, \tau))] + \mathbb{E}_q[\ln(p(\tau))] \tag{A32}$$

This way, the different elements of this equation can be calculated as:

$$\mathbb{E}_q[\ln(p(\mathbf{w}\,|\,\boldsymbol{\alpha}, \mathbf{a}))] = -\frac{D}{2}\ln(2\pi) + \frac{D}{2}\sum_{d=1}^{D}\left(\psi(a_{\mathrm{ff}_d}) - \ln(b_{\mathrm{ff}_d})\right)$$
$$- \sum_{d=1}^{D}\left(a_{\mathrm{ff}_d}\right) + b_0^\alpha \sum_{d=1}^{D}\left(\frac{a_{\mathrm{ff}_d}}{b_{\mathrm{ff}_d}}\right) \tag{A33}$$

$$\mathbb{E}_q[\ln(p(\boldsymbol{\alpha}))] = (a_0^\alpha \ln(b_0^\alpha) - \ln(\Gamma(a_0^\alpha)))$$
$$+ \sum_{d=1}^{D}\left(-b_0^\alpha \frac{a_{\mathrm{ff}_d}}{b_{\mathrm{ff}_d}} + (a_0^\alpha - 1)\left(\psi(a_{\mathrm{ff}_d}) - \ln(b_{\mathrm{ff}_d})\right)\right) \tag{A34}$$

$$\mathbb{E}_q[\ln(p(\mathbf{w}, \boldsymbol{\alpha}))] = \left(\frac{D}{2} + a_0^\alpha - 1\right)\sum_{d=1}^{D}\left(\psi(a_{\mathrm{ff}_d}) - \ln(b_{\mathrm{ff}_d})\right) - \frac{D}{2}\ln(2\pi)$$
$$+ (a_0^\alpha \ln(b_0^\alpha) - \ln(\Gamma(a_0^\alpha))) - \sum_{d=1}^{D}\left(a_{\mathrm{ff}_d}\right) \tag{A35}$$

$$\mathbb{E}_q[\ln(p(\boldsymbol{y}|\,\mathbf{a},\mathbf{X},\tau))] \;=\; -\frac{ND}{2}\ln(2\pi) + \frac{D}{2}\sum_{n=1}^{N}\left(\mathbb{E}_q[\ln(\tau)]\right)$$

$$-\frac{1}{2}\mathbb{E}_q[\tau]\left(\sum_{n=1}^{N}y_n^2 - 2\,\mathrm{Tr}\left\{\boldsymbol{y}^{\mathrm{T}}\,\mathbf{K}\langle\mathbf{a}\rangle\right\} + \mathrm{Tr}\left\{\langle\mathbf{a},\mathbf{a}^{\mathrm{T}}\rangle\,\mathbf{K}^{\mathrm{T}}\,\mathbf{K}\right\}\right)$$

$$=\; -\frac{ND}{2}\ln(2\pi) + \frac{D}{2}(\psi(a_\tau) - \ln(b_\tau)) - a_\tau + \frac{a_\tau}{b_\tau}b_0^\tau \tag{A36}$$

$$\mathbb{E}_q[\ln(p(\tau))] \;=\; a_0^\tau\ln(b_0^\tau) - \ln(\Gamma(a_0^\tau)) - b_0^\tau\frac{a_\tau}{b_\tau} + (a_0^\tau - 1)(\psi(a_\tau) - \ln(b_\tau)) \tag{A37}$$

$$\mathbb{E}_q[\ln(p(\boldsymbol{y},\tau|\,\mathbf{a},\mathbf{X}))] \;=\; -\frac{ND}{2}\ln(2\pi) - a_\tau + a_0^\tau\ln(b_0^\tau) - \ln(\Gamma(a_0^\tau)) +$$

$$\left(\frac{D}{2} + a_0^\tau - 1\right)(\psi(a_\tau) - \ln(b_\tau)) \tag{A38}$$

Appendix B.2. Terms of Entropy

The second term in the LB expression is the entropy of the model parameters:

$$\mathbb{E}_q[\ln(q(\Theta))] \;=\; \mathbb{E}_q[\ln(q(\mathbf{w}))] + \mathbb{E}_q[\ln(q(\boldsymbol{\alpha}))] + \mathbb{E}_q[\ln(q(\tau))], \tag{A39}$$

where the entropy of these parameters is:

$$\mathbb{E}_q[\ln(q(\mathbf{w}))] \;=\; \frac{D}{2}\ln(2\pi e) + \frac{D}{2}\ln|\Sigma_{\mathbf{w}}| \tag{A40}$$

$$\mathbb{E}_q\left[\ln\left(q\left(\boldsymbol{\alpha}^{(m)}\right)\right)\right] =$$

$$\sum_{k=1}^{K_c}\left(a_{\mathrm{ff}_k^{(m)}} + \ln\left(\Gamma\left(a_{\mathrm{ff}_k^{(m)}}\right)\right) - \left(1 - a_{\mathrm{ff}_k^{(m)}}\right)\psi\left(a_{\mathrm{ff}_k^{(m)}}\right) - \ln\left(b_{\mathrm{ff}_k^{(m)}}\right)\right) \tag{A41}$$

$$\mathbb{E}_q\left[\ln\left(q\left(\varnothing^{(m)}\right)\right)\right] \;=\; a_{\varnothing^{(m)}} + \ln(\Gamma(a_{\varnothing^{(m)}})) - (1 - a_{\varnothing^{(m)}})\psi(a_{\varnothing^{(m)}}) - \ln(b_{\varnothing^{(m)}}) \tag{A42}$$

Appendix B.3. Complete Lower Bound

Finally, joining Equations (A32) and (A39), the complete lower bound is calculated as:

$$L_q \;=\; -\left(\frac{D}{2} + a_0^\alpha - 1\right)\sum_{k=1}^{K_c}(\ln(b_{\alpha_k})) - \left(\frac{D}{2} + a_0^\tau - 1\right)(\ln(b_\tau))$$

$$-\frac{D}{2}\ln|\Sigma_{\mathbf{w}}| + \sum_{k=1}^{K_c}(\ln(b_{\alpha_k})) + \ln(b_\tau) + \mathrm{const} \tag{A43}$$

References

1. Carvalho, A.F.; Solmi, M.; Sanches, M.; Machado, M.O.; Stubbs, B.; Ajnakina, O.; Sherman, C.; Sun, Y.R.; Liu, C.S.; Brunoni, A.R.; et al. Evidence-based umbrella review of 162 peripheral biomarkers for major mental disorders. *Transl. Psychiatry* **2020**, *10*, 152. [CrossRef] [PubMed]
2. Widing, L.; Simonsen, C.; Flaaten, C.B.; Haatveit, B.; Vik, R.K.; Wold, K.F.; Åsbø, G.; Ueland, T.; Melle, I. Symptom Profiles in Psychotic Disorder Not Otherwise Specified. *Front. Psychiatry* **2020**, *11*, 580444. [CrossRef] [PubMed]
3. Correll, C.U.; Brevig, T.; Brain, C. Patient characteristics, burden and pharmacotherapy of treatment-resistant schizophrenia: Results from a survey of 204 US psychiatrists. *BMC Psychiatry* **2019**, *19*, 362. [CrossRef] [PubMed]

4. Roberts, L.W.; Chan, S.; Torous, J. New tests, new tools: Mobile and connected technologies in advancing psychiatric diagnosis. *NPJ Digit. Med.* **2018**, *1*, 20176. [CrossRef] [PubMed]

5. Li, Z.; Li, W.; Wei, Y.; Gui, G.; Zhang, R.; Liu, H.; Chen, Y.; Jiang, Y. Deep learning based automatic diagnosis of first-episode psychosis, bipolar disorder and healthy controls. *Comput. Med. Imaging Graph.* **2021**, *89*, 101882. [CrossRef] [PubMed]

6. Trakadis, Y.J.; Sardaar, S.; Chen, A.; Fulginiti, V.; Krishnan, A. Machine learning in schizophrenia genomics, a case-control study using 5090 exomes. *Am. J. Med. Genet. Part B Neuropsychiatr. Genet.* **2019**, *180*, 103–112. [CrossRef]

7. Hettige, N.C.; Nguyen, T.B.; Yuan, C.; Rajakulendran, T.; Baddour, J.; Bhagwat, N.; Bani-Fatemi, A.; Voineskos, A.N.; Chakravarty, M.M.; De Luca, V. Classification of suicide attempters in schizophrenia using sociocultural and clinical features: A machine learning approach. *Gen. Hosp. Psychiatry* **2017**, *47*, 20–28. [CrossRef]

8. Xiao, Y.; Yan, Z.; Zhao, Y.; Tao, B.; Sun, H.; Li, F.; Yao, L.; Zhang, W.; Chandan, S.; Liu, J.; et al. Support vector machine-based classification of first episode drug-naïve schizophrenia patients and healthy controls using structural MRI. *Schizophr. Res.* **2019**, *214*, 11–17. [CrossRef]

9. Guo, Y.; Qiu, J.; Lu, W. Support Vector Machine-Based Schizophrenia Classification Using Morphological Information from Amygdaloid and Hippocampal Subregions. *Brain Sci.* **2020**, *10*, 562. [CrossRef]

10. Jahmunah, V.; Oh, S.L.; Rajinikanth, V.; Ciaccio, E.J.; Cheong, K.H.; Arunkumar, N.; Acharya, U.R. Automated detection of schizophrenia using nonlinear signal processing methods. *Artif. Intell. Med.* **2019**, *100*, 101698. [CrossRef]

11. Brownlee, J. Recursive Feature Elimination (RFE) for Feature Selection in Python. Machine Learning Mastery. 2020. Available online: https://machinelearningmastery.com/rfe-feature-selection-in-python/ (accessed on 25 May 2020).

12. Breiman, L. Random forests. *Mach. Learn.* **2001**, *45*, 5–32. [CrossRef]

13. Li, X.; Chen, W.; Zhang, Q.; Wu, L. Building auto-encoder intrusion detection system based on random forest feature selection. *Comput. Secur.* **2020**, *95*, 101851. [CrossRef]

14. Zou, H.; Hastie, T. Regularization and variable selection via the elastic net. *J. R. Stat. Soc. Ser. (Stat. Methodol.)* **2005**, *67*, 301–320. [CrossRef]

15. Amini, F.; Hu, G. A two-layer feature selection method using genetic algorithm and elastic net. *Expert Syst. Appl.* **2021**, *166*, 114072. [CrossRef]

16. Shen, L.; Qi, Y.; Kim, S.; Nho, K.; Wan, J.; Risacher, S.L.; Saykin, A.J. Sparse bayesian learning for identifying imaging biomarkers in AD prediction. In Proceedings of the International Conference on Medical Image Computing and Computer-Assisted Intervention, Beijing, China, 20–24 September 2010; pp. 611–618.

17. Sabuncu, M.R.; Van Leemput, K. The relevance voxel machine (RVoxM): A self-tuning Bayesian model for informative image-based prediction. *IEEE Trans. Med. Imaging* **2012**, *31*, 2290–2306. [CrossRef]

18. Sabuncu, M.R. A sparse Bayesian learning algorithm for longitudinal image data. In Proceedings of the International Conference on Medical Image Computing and Computer-Assisted Intervention, Munich, Germany, 5–9 October 2015; pp. 411–418.

19. Parrado-Hernández, E.; Gómez-Verdejo, V.; Martínez-Ramón, M.; Shawe-Taylor, J.; Alonso, P.; Pujol, J.; Menchón, J.M.; Cardoner, N.; Soriano-Mas, C. Discovering brain regions relevant to obsessive–compulsive disorder identification through bagging and transduction. *Med. Image Anal.* **2014**, *18*, 435–448. [CrossRef]

20. Gómez-Verdejo, V.; Parrado-Hernández, E.; Tohka, J. Sign-consistency based variable importance for machine learning in brain imaging. *Neuroinformatics* **2019**, *17*, 593–609. [CrossRef]

21. Sevilla-Salcedo, C.; Gómez-Verdejo, V.; Tohka, J. Regularized Bagged Canonical Component Analysis for Multiclass Learning in Brain Imaging. *Neuroinformatics* **2020**, *18*, 641–659. [CrossRef]

22. Grimm, O.; Gass, N.; Weber-Fahr, W.; Sartorius, A.; Schenker, E.; Spedding, M.; Risterucci, C.; Schweiger, J.I.; Böhringer, A.; Zang, Z.; et al. Acute ketamine challenge increases resting state prefrontal-hippocampal connectivity in both humans and rats. *Psychopharmacology* **2015**, *232*, 4231–4241. [CrossRef]

23. Hadar, R.; Soto-Montenegro, M.L.; Götz, T.; Wieske, F.; Sohr, R.; Desco, M.; Hamani, C.; Weiner, I.; Pascau, J.; Winter, C. Using a maternal immune stimulation model of schizophrenia to study behavioral and neurobiological alterations over the developmental course. *Schizophr. Res.* **2015**, *166*, 238–247. [CrossRef]

24. Romero-Miguel, D.; Casquero-Veiga, M.; MacDowell, K.S.; Torres-Sanchez, S.; Garcia-Partida, J.A.; Lamanna-Rama, N.; Romero-Miranda, A.; Berrocoso, E.; Leza, J.C.; Desco, M.; et al. A Characterization of the Effects of Minocycline Treatment During Adolescence on Structural, Metabolic, and Oxidative Stress Parameters in a Maternal Immune Stimulation Model of Neurodevelopmental Brain Disorders. *Int. J. Neuropsychopharmacol.* **2021**, *24*, 734–748. [CrossRef] [PubMed]

25. Ozawa, K.; Hashimoto, K.; Kishimoto, T.; Shimizu, E.; Ishikura, H.; Iyo, M. Immune activation during pregnancy in mice leads to dopaminergic hyperfunction and cognitive impairment in the offspring: A neurodevelopmental animal model of schizophrenia. *Biol. Psychiatry* **2006**, *59*, 546–554. [CrossRef] [PubMed]

26. Zhu, F.; Zheng, Y.; Liu, Y.; Zhang, X.; Zhao, J. Minocycline alleviates behavioral deficits and inhibits microglial activation in the offspring of pregnant mice after administration of polyriboinosinic–polyribocytidilic acid. *Psychiatry Res.* **2014**, *219*, 680–686. [CrossRef] [PubMed]

27. Meyer, U.; Feldon, J. To poly (I: C) or not to poly (I: C): Advancing preclinical schizophrenia research through the use of prenatal immune activation models. *Neuropharmacology* **2012**, *62*, 1308–1321. [CrossRef]

28. Casquero-Veiga, M.; Garcia-Garcia, D.; MacDowell, K.S.; Perez-Caballero, L.; Torres-Sanchez, S.; Fraguas, D.; Berrocoso, E.; Leza, J.C.; Arango, C.; Desco, M.; et al. Risperidone administered during adolescence induced metabolic, anatomical and

inflammatory/oxidative changes in adult brain: A pet and mri study in the maternal immune stimulation animal model. *Eur. Neuropsychopharmacol.* **2019**, *29*, 880–896. [CrossRef]

29. Valdes Hernandez, P.A.; Sumiyoshi, A.; Nonaka, H.; Haga, R.; Aubert Vasquez, E.; Ogawa, T.; Iturria Medina, Y.; Riera, J.J.; Kawashima, R. An in vivo MRI template set for morphometry, tissue segmentation, and fMRI localization in rats. *Front. Neuroinform.* **2011**, *5*, 26.

30. Bishop, C.M. Bayesian PCA. In *Advances in Neural Information Processing Systems*; The MIT Press: Cambridge, MA, USA, 1999 ; pp. 382–388.

31. Klami, A.; Virtanen, S.; Kaski, S. Bayesian canonical correlation analysis. *J. Mach. Learn. Res.* **2013**, *14*, 965–1003.

32. Bishop, C.M. Pattern recognition. *Mach. Learn.* **2006**, *128* , 1–39 .

33. Schölkopf, B.; Herbrich, R.; Smola, A.J. A generalized representer theorem. In Proceedings of the International Conference on Computational Learning Theory, Amsterdam, The Netherlands, 16–19 July 2001; pp. 416–426.

34. Blei, D.M.; Kucukelbir, A.; McAuliffe, J.D. Variational Inference: A Review for Statisticians. *J. Am. Stat. Assoc.* **2017**, *112*, 859–877. doi: [CrossRef]

35. Rasmussen, C.E. Gaussian processes in machine learning. In *Summer School on Machine Learning*; Springer: Berlin/Heidelberg, Germany, 2003; pp. 63–71.

36. Steinwart, I.; Christmann, A. *Support Vector Machines*; Springer Science & Business Media: New York, NY, USA, 2008.

37. Pedregosa, F.; Varoquaux, G.; Gramfort, A.; Michel, V.; Thirion, B.; Grisel, O.; Blondel, M.; Prettenhofer, P.; Weiss, R.; Dubourg, V.; et al. Scikit-learn: Machine Learning in Python. *J. Mach. Learn. Res.* **2011**, *12*, 2825–2830.

38. Sevilla-Salcedo, C.; Gómez-Verdejo, V.; Olmos, P.M. Sparse semi-supervised heterogeneous interbattery bayesian analysis. *Pattern Recognit.* **2021**, *120*, 108141. [CrossRef]

39. Sevilla-Salcedo, C.; Guerrero-López, A.; Olmos, P.M.; Gómez-Verdejo, V. Bayesian Sparse Factor Analysis with Kernelized Observations. *arXiv* **2020**, arXiv:2006.00968.

40. Styner, M.; Lieberman, J.A.; McClure, R.K.; Weinberger, D.R.; Jones, D.W.; Gerig, G. Morphometric analysis of lateral ventricles in schizophrenia and healthy controls regarding genetic and disease-specific factors. *Proc. Natl. Acad. Sci. USA* **2005**, *102*, 4872–4877. [CrossRef] [PubMed]

41. Rapado-Castro, M.; Villar-Arenzana, M.; Janssen, J.; Fraguas, D.; Bombin, I.; Castro-Fornieles, J.; Mayoral, M.; González-Pinto, A.; de la Serna, E.; Parellada, M.; et al. Fronto-Parietal Gray Matter Volume Loss Is Associated with Decreased Working Memory Performance in Adolescents with a First Episode of Psychosis. *J. Clin. Med.* **2021**, *10*, 3929. [CrossRef]

42. Wen, D.; Wang, J.; Yao, G.; Liu, S.; Li, X.; Li, J.; Li, H.; Xu, Y. Abnormality of subcortical volume and resting functional connectivity in adolescents with early-onset and prodromal schizophrenia. *J. Psychiatr. Res.* **2021**, *140*, 282–288 . [CrossRef]

43. Guo, S.; Kendrick, K.M.; Zhang, J.; Broome, M.; Yu, R.; Liu, Z.; Feng, J. Brain-wide functional inter-hemispheric disconnection is a potential biomarker for schizophrenia and distinguishes it from depression. *Neuroimage Clin.* **2013**, *2*, 818–826. [CrossRef]

44. Boklage, C.E. Schizophrenia, brain asymmetry development, and twinning: Cellular relationship with etiological and possibly prognostic implications. *Biol. Psychiatry* **1977**, *12*, 19–35.

45. Casquero-Veiga, M.; Romero-Miguel, D.; MacDowell, K.S.; Torres-Sanchez, S.; Garcia-Partida, J.A.; Lamanna-Rama, N.; Gómez-Rangel, V.; Romero-Miranda, A.; Berrocoso, E.; Leza, J.C.; et al. Omega-3 fatty acids during adolescence prevent schizophrenia-related behavioural deficits: Neurophysiological evidences from the prenatal viral infection with PolyI: C. *Eur. Neuropsychopharmacol.* **2021**, *46*, 14–27. [CrossRef]

46. Bortz, D.M.; Grace, A.A. Medial septum activation produces opposite effects on dopamine neuron activity in the ventral tegmental area and substantia nigra in MAM vs. normal rats. *NPJ Schizophr.* **2018**, *4*, 17. [CrossRef]

47. Takeuchi, Y.; Nagy, A.; Barcsai, L.; Li, Q.; Ohsawa, M.; Mizuseki, K.; Berényi, A. The medial septum as a potential target for treating brain disorders associated with oscillopathies. *Front. Neural Circuits* **2021**, *15*, 701080. [CrossRef] [PubMed]

48. McGlinchey, E.M.; Aston-Jones, G. Dorsal hippocampus drives context-induced cocaine seeking via inputs to lateral septum. *Neuropsychopharmacology* **2018**, *43*, 987–1000. [CrossRef] [PubMed]

49. Pantazis, C.B.; Aston-Jones, G. Lateral septum inhibition reduces motivation for cocaine: Reversal by diazepam. *Addict. Biol.* **2020**, *25*, e12742. [CrossRef] [PubMed]

50. Gárate-Pérez, M.F.; Méndez, A.; Bahamondes, C.; Sanhueza, C.; Guzmán, F.; Reyes-Parada, M.; Sotomayor-Zárate, R.; Renard, G.M. Vasopressin in the lateral septum decreases conditioned place preference to amphetamine and nucleus accumbens dopamine release. *Addict. Biol.* **2021**, *26*, e12851. [CrossRef] [PubMed]

51. Yang, M.; Gao, S.; Zhang, X. Cognitive deficits and white matter abnormalities in never-treated first-episode schizophrenia. *Transl. Psychiatry* **2020**, *10*, 368. [CrossRef]

52. Kim, S.E.; Jung, S.; Sung, G.; Bang, M.; Lee, S.H. Impaired cerebro-cerebellar white matter connectivity and its associations with cognitive function in patients with schizophrenia. *NPJ Schizophr.* **2021**, *7*, 38. [CrossRef]

applied sciences

Article

All You Need Is a Few Dots to Label CT Images for Organ Segmentation

Mingeon Ju [†], Moonhyun Lee [†], Jaeyoung Lee [†], Jaewoo Yang [†], Seunghan Yoon [†] and Younghoon Kim [*]

Major in Bio Artificial Intelligence, Hanyang University at Ansan, Ansan 15588, Korea; msgee@hanyang.ac.kr (M.J.); greenzip0510@hanyang.ac.kr (M.L.); wayexists02@hanyang.ac.kr (J.L.); onnoo@hanyang.ac.kr (J.Y.); shyoon93@hanyang.ac.kr (S.Y.)
* Correspondence: nongaussian@hanyang.ac.kr
† These authors contributed equally to this work.

Abstract: Image segmentation is used to analyze medical images quantitatively for diagnosis and treatment planning. Since manual segmentation requires considerable time and effort from experts, research to automatically perform segmentation is in progress. Recent studies using deep learning have improved performance but need many labeled data. Although there are public datasets for research, manual labeling is required in an area where labeling is not performed to train a model. We propose a deep-learning-based tool that can easily create training data to alleviate this inconvenience. The proposed tool receives a CT image and the pixels of organs the user wants to segment as inputs and extract the features of the CT image using a deep learning network. Then, pixels that have similar features are classified to the identical organ. The advantage of the proposed tool is that it can be trained with a small number of labeled data. After training with 25 labeled CT images, our tool shows competitive results when it is compared to the state-of-the-art segmentation algorithms, such as UNet and DeepNetV3.

Keywords: medical image segmentation; CT image segmentation; deep learning; kernel density; semi-automated labeling tool

Citation: Ju, M.; Lee, M.; Lee, J.; Yang, J.; Yoon, S.; Kim, Y. All You Need Is a Few Dots to Label CT Images for Organ Segmentation. *Appl. Sci.* **2022**, *12*, 1328. https://doi.org/10.3390/app12031328

Academic Editor: Jan Egger

Received: 31 December 2021
Accepted: 25 January 2022
Published: 26 January 2022

Publisher's Note: MDPI stays neutral with regard to jurisdictional claims in published maps and institutional affiliations.

1. Introduction

Image segmentation is used to analyze the medical image quantitatively for the quantification of tissue volume, diagnosis, treatment planning, and computer-integrated surgery [1]. However, it takes a lot of time and effort for a radiologist and doctor to conduct segmentation on CT images of each patient. Therefore, the necessity for technology that can accelerate segmentation on CT images has been highlighted consistently.

Over the past few decades, CNN-based deep learning techniques have made remarkable success in computer vision tasks, and there have been attempts to apply CNN-based methods in the field of medical image segmentation [2–6]. Afterward, UNet [7], which has the U-shaped encoder-decoder architecture with skip-connection, significantly improved the performance of the medical image segmentation task. Since the introduction of UNet, many variants using its architecture have been studied for segmentation tasks on organs and muscles of CT images in the medical field.

Although many deep learning models perform well enough to generate segmentation on CT images, it is not easy for medical institutions to train a network fit for their purpose. First, making public labeled data is difficult because the privacy of local data in medical institutions is essential. In addition, it cannot be guaranteed that the network trained by public data will conduct segmentation well on data owned by each institution since the distribution of pixels in the CT image differs depending on radiography equipment. Finally, the tedious work of labeling is needed whenever a segmentation task is conducted on an unlabeled organ in a public dataset.

In this research, we propose a deep-learning-based tool that can help with the labeling task, which is required to make the labeled data for a network performing segmentation and reduce the time and effort of experts. The network used for our tool can be trained with a small number of labeled data. After training, the tool shows the region of the organ when the user puts a dot on the target organ to be segmented. The user can conveniently adjust the threshold so that the labeling can be performed flexibly according to the data distribution. We expect that the labeled data for the medical image segmentation model would be easily made using our proposed labeling tool.

We compared our tool with two representative segmentation methods, UNet and DeepLabV3, for evaluation. Our technique is challenging to compare with the automated segmentation methods because a user has to annotate the pixels and set the threshold. Therefore, we proceed with two preceding experiments to determine the number of input anchors and the threshold for the experiment. For an efficient experiment, the anchors are extracted from the ground truth. Finally, Dice Score and Hausdorff Distance are used to measure the performance.

In this paper, our main contributions are as follows:

- We propose a novel labeling tool that can be trained with only a few labeled data by incorporating visual features and locality information extracted from Feature Encoder and Gaussian kernel.
- We utilize anchor pixels from user interactions for easily segmenting organs. The pixels can be selected anywhere on the target organs, with no additional constraints, such as annotating bounding-box, extreme points.
- Our tool provides an additional function to refine the segmentation result in detail by modifying the threshold value.

The following section discusses recent deep-learning-based segmentation models, interactive medical image segmentation models, and differences between the proposed method and existing interactive segmentation models. Section 3 introduces the proposed method, the dataset used for training, the metric for performance evaluation, and the application that can conveniently perform segmentation using our model. The results of measuring the performance of our method are presented in Section 4. Finally, Section 5 summarizes the proposed method and discusses future research directions.

2. Related Work

In this section, we study the recent work for image segmentation, which is categorized into UNet, Transformer-based models, and interactive segmentation methods. While many recent segmentation algorithms have adopted UNet and Transformer, they require a large set of fully labeled segmentation images. In this work, we exploit both UNet-based model architecture and an interactive segmentation strategy to generate a segmentation-labeled image with little effort.

2.1. Deep-Learning-Based Medical Image Segmentation

After the success of UNet on medical image segmentation tasks, there have been many studies trying to utilize its architecture or applying additional techniques to improve the performance [8–11]. For example, UNet combined with self-attention gates [8] on coarsely extracted feature maps from the encoding path is used for multi-organ segmentation. This gating method relieved the noisy response in skip-connection, capturing more precise textural and structural information on the input image patches. Bottleneck Feature supervised BS-UNet [9] connects its network with two UNets for liver and tumor segmentation. The first network is UNet with no skip-connection, which works as an auto-encoder learning the feature information from labeled data. The second network is the original UNet that performs the segmentation task. GIU-Net [12] utilize a graph-cut algorithm for multi-organ segmentation (skip-connection), which modified the structure of the skip-connection and used it to solve the task. Targeting organs-at-risk (OARs) is also crucial work in medical facilities. In [11], the convolution layers of UNet are replaced in the context aggregation block to learn from wide-range features to perform segmentation of OARs in cervical cancer.

Recently, research to improve medical image segmentation performance is still in progress. For instance, some studies adapt the UNet architecture as the backbone and exploit other deep learning techniques. UTNet [13] utilize a transformer encoder block and a decoder block as the last part of each convolution layer. Ref. [14] extracts fine-grained features by utilizing various sizes of the receptive field. It inserted a channel attention (CA) block into skip connection and applied a hybrid dilated attention convolutional (HDAC) layer to the last encoder output. X-Net [15] proposes an X-shaped network that can learn in parallel using two different branches: the CNN branch and transformer branch. As the Vision Transformer rises in various fields of the vision task, many studies utilizing vision transformers as the backbone have been conducted for medical image segmentation. In [16], a U-shaped network with a vision transformer base encoder for brain tumors and abdominal organ segmentation of 3D image patches is proposed, and a network consisting of pure transformers is introduced in [17]. The authors, inspired by Swin Transformer [18] and UNet, built an encoder-decoder architecture composed of only transformers using Swin Transformer blocks.

2.2. Interactive Medical Image Segmentation

Hence, interactive segmentation tools have been studied, which can assist in generating segmentation datasets on CT images with little effort. For instance, landmark points [19] or bounding boxes [20] could be guides for segmentation. Nevertheless, these studies still require many labeled data or careful guidance, which can make data collection expensive, especially in the medical field. Therefore, to solve such drawbacks, we propose an interactive labeling tool that can be helpful in the medical field, making it possible to gather data with ease and less time.

In addition, the performance of active contour models is also improved. Ref. [21] proposed an additive bias correction model to reduce the long execution time of the existing bias correction model. Since these active contour models must designate the region of interest (ROI) and detect the object's boundary, the segmentation result can significantly differ depending on the given ROI. Our study focused on users conveniently performing segmentation on abdominal CT images.

3. Materials and Methods

We now define the dataset and propose our segmentation tool based on a deep neural network. Initially, we explain the CT image dataset used in this paper. Next, we introduce our model architecture consisting of a feature encoder and two kernel modules, the used metrics, and the loss function. Finally, we show our application implemented with these materials.

3.1. Dataset

We utilize the BTCV [22] dataset (https://www.synapse.org/#!Synapse:syn3193805/files/, accessed on 1 September 2021). The dataset has 13 organs with segmentation performed by a trained person and reviewed by a radiologist. In this study, we measured the performance through four organs: liver, left kidney, right kidney, and spleen, located in the abdomen. In Figure 1, we provide abdominal CT images of four different people; each organ has its colors and textures, and the positions are relatively similar.

To build our dataset, we sampled 202 CT images that have 4 organs: liver, spleen, left kidney, and right kidney. Among them, only 25 CT images were used for training and 177 CT images for testing. All CT images have segmentation labels. We exploited segmentation labels of the training set to randomly sample anchor pixels instead of manually picking them. For preprocessing, we arranged the range of CT image pixels to $[-1, 1]$ and applied histogram equalization.

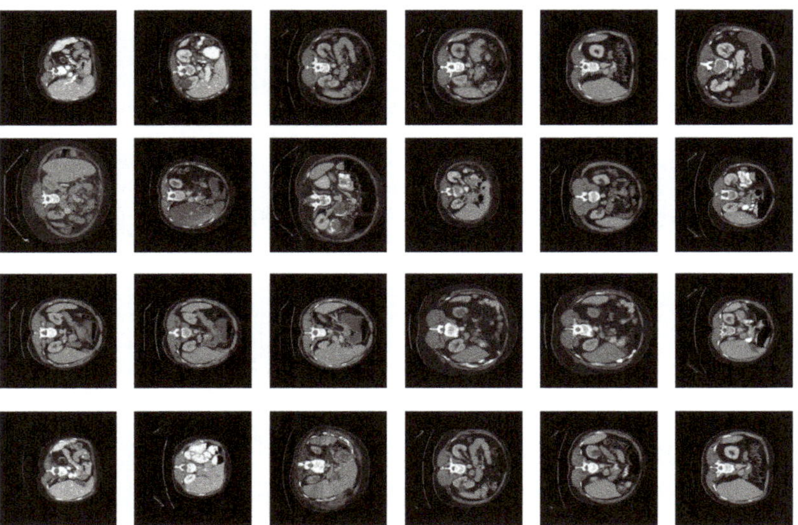

Figure 1. CT image examples that show different colors and textures for each organs.

3.2. Proposed Method

We propose a semi-automated tool that generates organ segmentation. Given a CT image, the radiologists annotate a few pixels—an anchor to specify the target organ. The pixels near the *anchor pixels*, which have similar characteristics, are candidates of the target organ to be segmented. To capture the segmentation region of the target organs, we utilize two types of information: *visual features* and *locality*. In Figure 2, we provide a schematic representation of our network. Our proposed model consists of two sub-modules: Feature Encoder $\mathcal{F}$ and Kernel Function $\mathcal{K}$. The Feature Encoder $\mathcal{F}$ extracts the feature vector for each pixel, which is used to calculate the visual similarity between pixels. We exploit a modified U-Net architecture to implement the Feature Encoder $\mathcal{F}$.

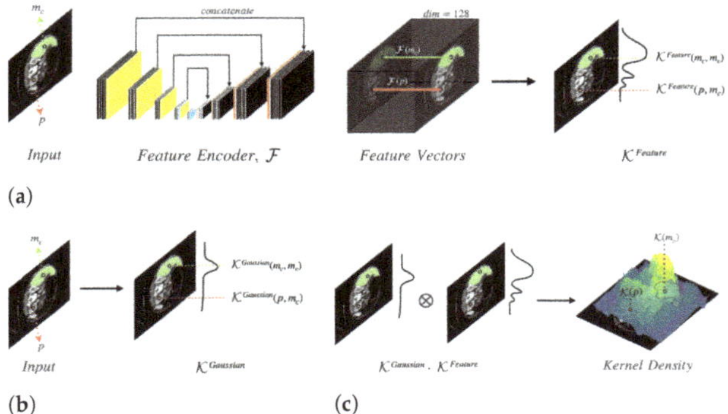

Figure 2. Model architecture that consists of 2 kernels. These kernels are used to compute the kernel density, which has a high value in the region of the target organ: (**a**) Feature similarity kernel density; (**b**) Gaussian kernel density; (**c**) Kernel density.

The Kernel Function $\mathcal{K}$ is used to capture both *visual features* and *locality* information. In a CT image, let M_c be a set of anchor pixels that a user picks on the c-th organ. The Kernel Function $\mathcal{K}$ computes the density for all pixels, which represents visual and positional similarity. As illustrated in Figure 2c, the *kernel density* is made of two types of densities:

feature similarity kernel density $\mathcal{K}_c^{\text{Feature}}$ (shown in Figure 2a) and Gaussian kernel density $\mathcal{K}_c^{\text{Gaussian}}$ (shown in Figure 2b). The *kernel density* $\mathcal{K}_c$ of a pixel p is computed as follows:

$$\mathcal{K}_c(p) = \sum_{m_c \in \mathbf{M}_c} \mathcal{K}_c^{\text{Feature}}(p, m_c) \mathcal{K}_c^{\text{Gaussian}}(p, m_c) \tag{1}$$

where p is an arbitrary pixel, and m_c is an anchor pixel of the c-th organ. $\mathcal{K}_c^{\text{Feature}}$ means the feature similarity kernel density for the c-th organ. The feature similarity kernel density measures the feature similarity of two pixels. $\mathcal{K}_c^{\text{Gaussian}}$ indicates the Gaussian kernel density for the c-th organ. The Gaussian kernel density calculates the positional similarity of two pixels.

The kernel density has a maximum value at each anchor pixel m_c. In the other case, the density value becomes high if a pixel p is closer to the anchor pixels $\mathbf{M}_c$ and has a similar texture or color too. Therefore, we can figure out the region of the c-th organ, filtering pixels with a high density.

If a pixel p has similar visual features to anchor pixels M_c, the pixel p is more likely to belong to the c-th organ than other pixels. Using feature vectors $\mathcal{F}(p)$ for the pixel p and $\mathcal{F}(m_c)$ for an anchor pixel m_c, we can calculate the feature similarity kernel density by using the dot product between $\mathcal{F}(p)$ and $\mathcal{F}(m_c)$ as follows:

$$\mathcal{K}_c^{\text{Feature}}(p, m_c) = \mathcal{F}(p)^T \cdot \mathcal{F}(m_c) \tag{2}$$

where p is an arbitrary pixel, and m_c is an anchor pixel in the c-th organ. The operator $\cdot$ means dot product. If two pixels p and m_c have similar features, the feature similarity kernel density has high value.

Gaussian kernel quantified how close the pixel is to anchor points, that is, a pixel closer to anchor pixels has a higher density value. This density is required not to capture regions that have similar texture but are far from anchor points (in another organ). A pixel p should be closer to anchor pixels for being classified as a target label. The Gaussian kernel density is calculated as follows:

$$\mathcal{K}_c^{\text{Gaussian}}(p, m_c) = \exp(-\gamma_c ||p - m_c||_2^2) \tag{3}$$

where γ_c is a trainable precision parameter used in Gaussian kernel for the c-th organ, and $|| \cdot ||_2$ indicates L_2 norm.

In practice, we exploit a modified UNet architecture used as the Feature Encoder. We replace the convolution layers with the same padded convolution layers to fit the output resolution to the input resolution. We changed the output feature map channel to 128, which is the length of the feature vector of a pixel. Lastly, we utilize the Adam optimizer [23] to train our model.

3.3. Loss Function

We utilize Noise Contrastive Estimation (NCE) [24] loss to build the objective function. NCE loss uses negative sampling to optimize model parameters by forcing the density of the positive sample to be high and the density of the negative sample to be low. Let $\mathbf{P}_c^{pos}$ be a set of pixels in the c-th organ and $\mathbf{P}_c^{neg}$ be a set of pixels not in the c-th organ. We collected positive pixels p_c^{pos} by randomly sampling from $\mathbf{P}_c^{pos}$ and collected negative pixels p_c^{neg} in the same way from $\mathbf{P}_c^{neg}$. Using these pixels, we formulate the NCE loss, whose form is from InfoNCE loss [25], as follows:

$$\mathcal{L}_{NCE} = \sum_c \mathcal{L}(c) \tag{4}$$

$$\text{where } \mathcal{L}(c) = \mathbb{E}_{p_c^{pos} \sim \mathbf{P}_c^{pos}} \left[-\log \frac{\exp(\mathcal{K}_c(p_c^{pos}))}{\mathbb{E}_{p_c^{neg} \sim \mathbf{P}_c^{neg}} \left[\exp(\mathcal{K}_c(p_c^{neg})) \right]} \right] \tag{5}$$

where $\mathcal{L}_c$ is the NCE loss for the c-th organ, p_c^{pos} is a pixel in the c-th organ, and p_c^{neg} is a pixel not in the c-th organ. $\mathcal{K}_c(\cdot)$ indicates the kernel density of a pixel, and $\mathbb{E}[\cdot]$ means the expectation.

3.4. Evaluation Metrics

To evaluate the performance of our model, we compare it with two baseline models: UNet and DeepNet-V3. We benchmark the segmentation performance with two segmentation measures: dice score and Hausdorff distance. Given the predicted region P and ground truth region G, the dice score is computed as follows:

$$DICE = \frac{2(P \cap G)}{|P| + |G|} \tag{6}$$

where $|P|$ and $|G|$ mean the cardinality of P and G, respectively. The dice score indicates overlapped area between the segmentation prediction and ground truth, penalizing the missing prediction. In case the model prediction is more similar to ground truth in terms of segmentation size, the dice score is more highly evaluated. The Hausdorff distance is computed as follows:

$$HD(P, G) = \max(\max_{p \in P} \min_{g \in G} \| p - g \|, \max_{g \in G} \min_{p \in P} \| g - p \|) \tag{7}$$

In the inner terms, the first term calculates the maximum distance to the minimum distance from the predicted pixel to the ground truth pixel, and the second term calculates the distance. As a result, the Hausdorff distance represents the maximum distance between two pixels and that each pixel belongs to the exclusive region, respectively.

3.5. The Application Generating Label

After training the proposed method with a little training data, a user can perform organ segmentation on a CT image. Figure 3 shows our tool in use. It contains the input CT image, anchor pixels entered by a user, threshold value, and segmented target organ image. A user can create labels using our tool as following order:

1. Open a CT image to perform segmentation;
2. Select the target organ in category;
3. Pick anchor pixels on the target organ. Once anchor pixels are picked, proposed model computes kernel density for all pixels;
4. Extract the segmentation label by filtering the density that is higher than a threshold value.

To infer segmentation from a kernel density, a threshold value determines the lower density bound of the segmentation label. The suitable threshold value can be different for each organ since each organ has various sizes and shapes in a CT image. In order to enable adjusting the threshold value, we added a slider in the bottom of the interface, as illustrated Figure 3. Through the slider, a user can manipulate the threshold to generate a proper segmentation label on input CT image.

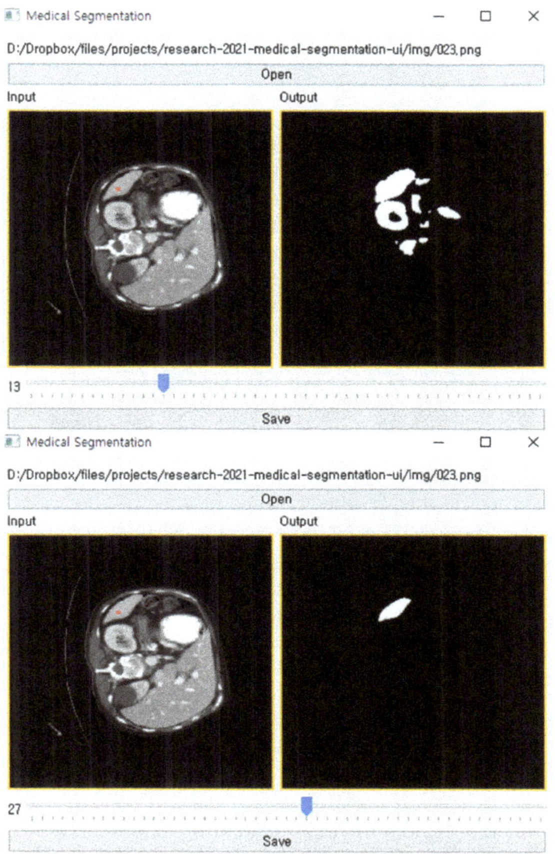

Figure 3. The user interface of our tool for reconstructing a segmentation label from a few anchor pixels. The upper figure shows the segmentation result when the threshold is 18. The lower figure is same as upper figure except for threshold value (which is 28).

4. Performance Tests

This section explains the experiment environment and results. This work was implemented with PyTorch [26], and the user interface was built with PyQT. When implementing evaluation metrics, we utilized the MONAI (https://monai.io/, accessed on 1 September 2021) library. The model training and inference are conducted on an Ubuntu machine with NVIDIA GTX 1080 Ti GPU installed. We visualized our experiments using the Weight & Bias [27] tool.

4.1. Performance Evaluation with Varying Threshold

A threshold value is a hyperparameter to draw segmentation results from a kernel density. Because the quality of the result depends on the threshold, the appropriate threshold value is required. To show the influence of the threshold value, we visualize segmentation images and measure the metrics of the segmentation, varying the threshold values from 0 to 50.

Given the anchor pixels and threshold value, Figure 4 illustrates segmentation images of each target organ. As a result, the lower threshold captures the region roughly, and the higher value shrinks the area to narrow. The dice score and Hausdorff distance are for measuring segmentation performance. As shown in Figure 5, the dice score reached

0.81 when the threshold value was 18, and the Hausdorff distance achieved a minimum distance of 11.96 when the threshold value was 33.

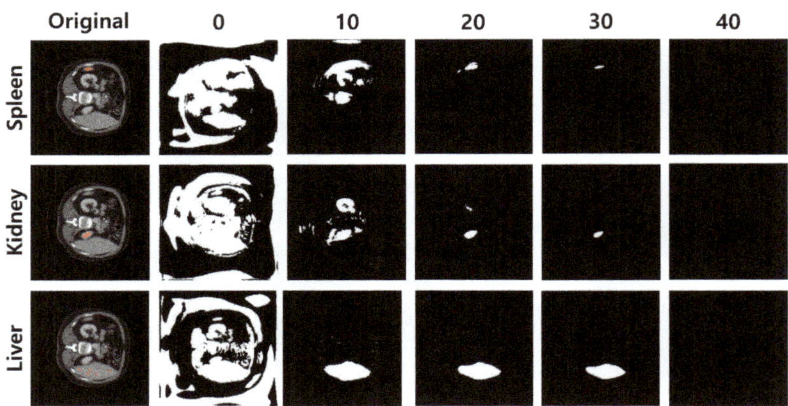

Figure 4. Segmentation Examples by Threshold. Each row represents each organ. First column shows original CT images, and from the second to sixth columns, the segmentation generated from 5 anchor pixels is shown. We visualize segmentation by different thresholds.

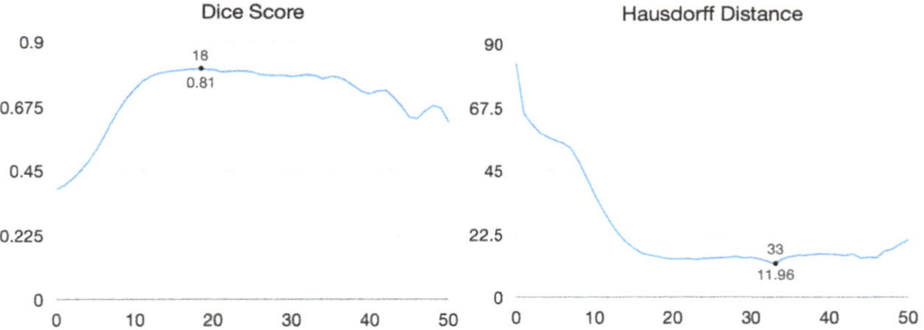

Figure 5. Dice score and Hausdorff distance by varying thresholds. Left graph demonstrates change in dice score by changing thresholds. Right graph shows Hausdorff distance by varying thresholds.

We analyzed that the dice score tends to be high when there is an intersection region between the predicted region and ground truth, even though a non-organ region is captured. In contrast, the Hausdorff distance tends to be low when the boundary of the predicted region is close to the ground truth. Our experiments focus more on capturing all regions of the target organ than predicting its boundary. As a result, we fix the threshold value as 18 in the following experiments.

4.2. Performance Evaluation with Varying the Number of Anchor Pixels

The proposed model requires a few anchor pixels on the organ to be segmented. The performance of the model can be affected by the number of anchor pixels picked by the user. The anchors are sampled randomly in the ground truth for convenience in the experiments. We tried to find out the best number of anchor pixels for our method.

As shown in Table 1, we measured the dice score and Hausdorff distance of our model according to the number of anchor pixels for the proposed method. The dice score is the lowest when using only one anchor pixel for each organ, and the dice score when using two or more anchor pixels is almost similar. However, the Hausdorff distance instead increases because other organs with similar textures to the target organ are often segmented.

Table 1. Dice score and Hausdorff distance by changing the number of anchors.

	Number of Anchor Pixels per Organs				
Model	1	2	3	4	5
DICE SCORE	0.717	0.752	0.761	0.762	0.759
Hausdorff Distance	29.155	14.956	18.748	20.433	17.951

Figure 6 shows how the number of anchor pixels affects the segmentation reconstructed from anchor pixels. As those figures illustrate, our model can find a segmentation region with a missing area when we pick only one anchor pixel. However, as the number of anchor pixels increases, the proposed method captured a more accurate region.

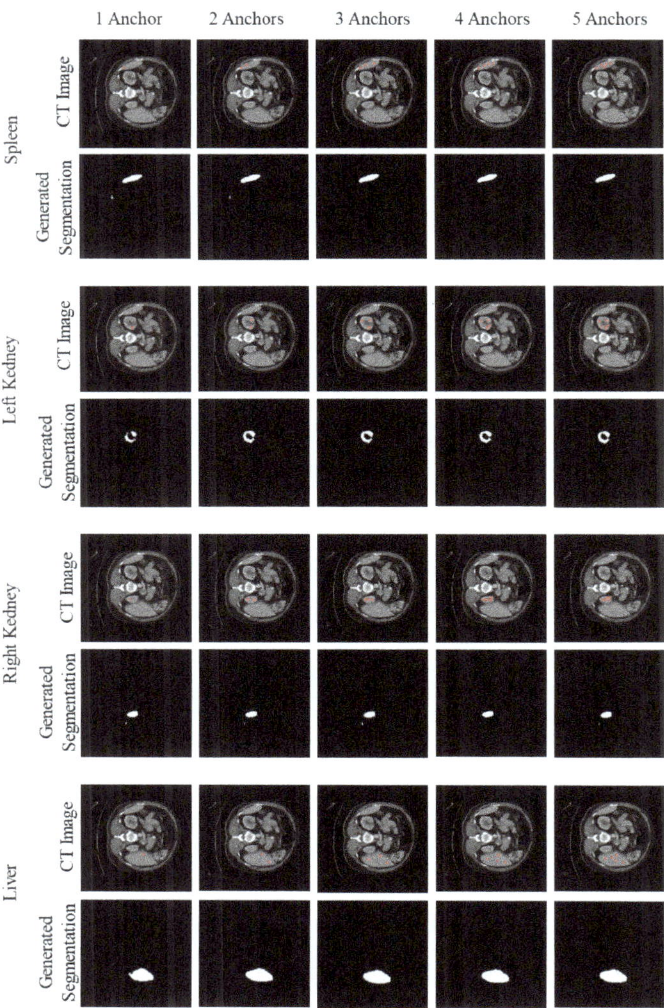

Figure 6. Case study to see how the generated segmentation is changed by a changing number of anchor pixels. The first and third columns are original CT images with some anchor pixels (red points). The second and fourth columns are generated segmentation using given anchor pixels.

Generally, more anchor pixels give better segmentation than fewer anchor pixels. However, we could find some exceptional cases that fewer anchor pixels show better performance. We analyzed that the position of anchor pixels also affects the performance. For instance, even if we have only one anchor pixel, the segmentation prediction is more accurate when the anchor pixel exists around the center of the organ. If we have more than one anchor pixel, but they are in the boundary, our model tends to capture only a part of the region. Since we automated the process of picking anchor pixels by randomly sampling them from pixels of the organ, the case that some anchor pixels cover the same region or exist in the boundary region frequently occurred.

4.3. Tests On Segmentation Accuracy

In our case, we focus on making the model show acceptable performance when a few labeled data are given. To verify the performance of our model, we measured the evaluation metrics by varying the number of training data. The x-axis of Figure 7 represents the number of people extracted from labeled images. In this experiment, we used only five CT images per person.

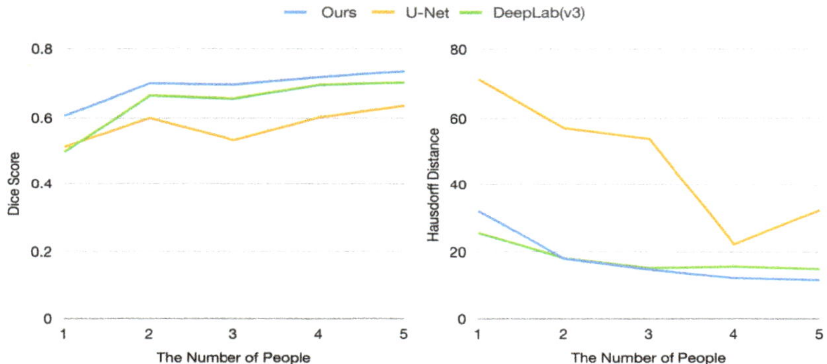

Figure 7. Dice score and Hausdorff distance by changing size of training data (number of people in training data).

Figure 7 shows the dice score and Hausdorff distance of the proposed model and baseline models. As illustrated in Figure 7, our model is superior to baseline models in terms of dice score. In particular, our model achieved a 0.6 dice score with training data of only one person, whereas baseline models achieved only about 0.5 dice score.

However, our model has a higher Hausdorff distance than DeepLab-V3 because segmentation generated by DeepLab-V3 is rarely scattered. In contrast, our method can generate segmentation with some scattered points since our approach exploits the density computed using visual similarity between pixels. If there are points that have similar features with anchor pixels, these points can be classified as part of the target organ, even if these pixels are apart from the target organ (this circumstance can also be found in Figure 4). This drawback can be relieved by carefully picking anchor pixels and thresholding.

We also compared our model with other semi-automatic algorithms, such as the active contour algorithm and the watershed algorithm. We exploited the true segmentation label to construct informative seed for the watershed algorithm. However, we found that these algorithms poorly detected the region of organs and achieved a dice score too low (0.04 and 0.07, respectively).

4.4. Ablation Study

This experiment explains the ablation study explain why these two kernel modules are necessary. As shown in Table 2, if we have only the Gaussian kernel density, the performance becomes lower than any other models in terms of both dice score and Hausdorff distance. At this point, we adjusted the threshold to 0.5 since the Gaussian kernel density generally has a small value. The threshold value of 0.5 is selected by the experiment to perform better. The performance drop is not huge when we take only the feature similarity kernel density (feature-only model). However, as illustrated in Figure 8, the feature-only model generates more scattered segmentation than ours. Therefore, utilizing the Gaussian kernel density to generate segmentation labels is necessary in order to avoid scattering, although the feature-only model has a tiny performance drop.

Table 2. Ablation study for kernels.

Gaussian Kernel	Feature Kernel	Dice Score	Hausdorff Distance
O	O	0.7590	23.31
O	X	0.3795	31.08
X	O	0.7370	23.85

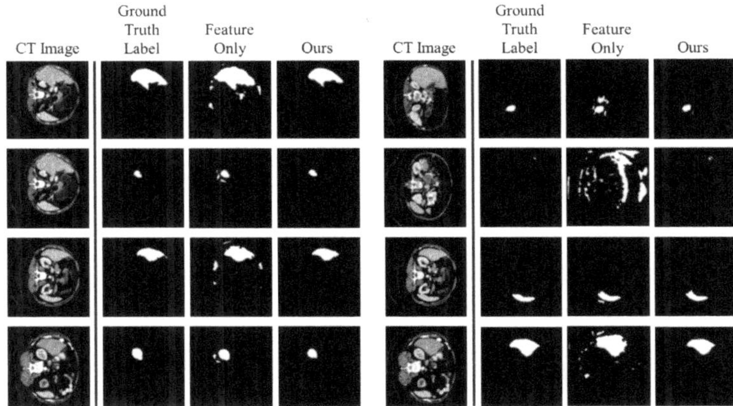

Figure 8. Ablation Study: Importance of the Gaussian Kernel. The first and fifth columns are CT images. The second and sixth columns are ground truth segmentation labels. The third and seventh columns are predictions generated by the model without the Gaussian kernel. The fourth and last columns are predictions generated by our model (Gaussian kernel + feature similarity kernel).

5. Conclusions

In this work, we proposed a tool that allows users to obtain segmentation labels by picking a few pixel points on the organ. The tool cannot segment without giving a pixel point but has the advantage that learning requires less labeled data than previous studies and makes data collection easier. This tool calculates the kernel density for all pixels, finds areas equal to or greater than a specific threshold value, and demonstrates its performance through experiments. However, the proposed method has a constraint that some non-target regions nearby a target organ could be classified as the same organ when the pixels have similar visual characteristics to a target organ. Furthermore, if an organ has various textures or multiple regions, our algorithm may miss some regions of the organ. Even though these constraints can be mitigated by carefully adjusting the threshold value, we should improve our algorithm in future work.

Author Contributions: Conceptualization, Y.K. and J.L.; methodology, Y.K. and J.L.; software, J.L. and J.Y.; validation, J.L. and J.Y.; formal analysis, M.J.; investigation, S.Y.; data curation, J.Y. and M.J.; writing—original draft preparation, S.Y. and M.L.; writing—review and editing, all authors; visualization, J.L.; supervision, Y.K.; project administration, Y.K.; funding acquisition, Y.K. All authors have read and agreed to the published version of the manuscript.

Funding: This work was supported in part by the Institute of Information and Communications Technology Planning and Evaluation (IITP) funded by the Korea Government (Ministry of Science and ICT (MSIT)) through the Artificial Intelligence Convergence Research Center (Hanyang University ERICA), under Grant 2020-0-01343, and in part by the National Research Foundation of Korea (NRF) funded by the Korea Government (MSIT) under Grant 2020R1G1A1011471.

Institutional Review Board Statement: Not applicable.

Informed Consent Statement: Not applicable.

Data Availability Statement: Not applicable.

Conflicts of Interest: The authors declare no conflict of interest. The funders had no role in the design of the study; in the collection, analyses, or interpretation of data; in the writing of the manuscript, or in the decision to publish the results.

References

1. Pham, D.L.; Xu, C.; Prince, J.L. Current methods in medical image segmentation. *Annu. Rev. Biomed. Eng.* **2000**, *2*, 315–337. [CrossRef]
2. Ciresan, D.; Giusti, A.; Gambardella, L.; Schmidhuber, J. Deep neural networks segment neuronal membranes in electron microscopy images. *Adv. Neural Inf. Process. Syst.* **2012**, *25*, 2843–2851.
3. Havaei, M.; Davy, A.; Warde-Farley, D.; Biard, A.; Courville, A.; Bengio, Y.; Pal, C.; Jodoin, P.M.; Larochelle, H. Brain tumor segmentation with deep neural networks. *Med. Image Anal.* **2017**, *35*, 18–31. [CrossRef] [PubMed]
4. Zhang, W.; Li, R.; Deng, H.; Wang, L.; Lin, W.; Ji, S.; Shen, D. Deep convolutional neural networks for multi-modality isointense infant brain image segmentation. *NeuroImage* **2015**, *108*, 214–224. [CrossRef] [PubMed]
5. Roth, H.R.; Farag, A.; Lu, L.; Turkbey, E.B.; Summers, R.M. Deep convolutional networks for pancreas segmentation in CT imaging. In *Proceedings of the Medical Imaging 2015: Image Processing*; International Society for Optics and Photonics: San Diego, CA, USA, 2015; Volume 9413, p. 94131G.
6. Tajbakhsh, N.; Shin, J.Y.; Gurudu, S.R.; Hurst, R.T.; Kendall, C.B.; Gotway, M.B.; Liang, J. Convolutional neural networks for medical image analysis: Full training or fine tuning? *IEEE Trans. Med. Imaging* **2016**, *35*, 1299–1312. [CrossRef] [PubMed]
7. Ronneberger, O.; Fischer, P.; Brox, T. *U-Net: Convolutional Networks for Biomedical Image Segmentation*; MICCAI: Strasbourg, France, 2015.
8. Liu, Y.; Lei, Y.; Fu, Y.; Wang, T.; Tang, X.; Jiang, X.; Curran, W.J.; Liu, T.; Patel, P.R.; Yang, X. CT-based Multi-organ Segmentation using a 3D Self-attention U-Net Network for Pancreatic Radiotherapy. *Med. Phys.* **2020**, *47*, 4316–4324. [CrossRef] [PubMed]
9. Li, S.; Tso, G.K.F.; He, K. Bottleneck feature supervised U-Net for pixel-wise liver and tumor segmentation. *Expert Syst. Appl.* **2020**, *145*, 113131. [CrossRef]
10. Wang, L.; Wang, B.; Xu, Z. Tumor Segmentation Based on Deeply Supervised Multi-Scale U-Net. In Proceedings of the 2019 IEEE International Conference on Bioinformatics and Biomedicine (BIBM), San Diego, CA, USA, 18–21 November 2019; pp. 746–749.
11. Liu, Z.; Liu, X.; Xiao, B.; Wang, S.; Miao, Z.; Sun, Y.; Zhang, F. Segmentation of organs-at-risk in cervical cancer CT images with a convolutional neural network. *Phys. Med. PM Int. J. Devoted Appl. Phys. Med. Biol. Off. J. Ital. Assoc. Biomed. Phys.* **2020**, *69*, 184–191. [CrossRef] [PubMed]
12. Liu, Z.; Song, Y.Q.; Sheng, V.S.; Wang, L.; Jiang, R.; Zhang, X.; Yuan, D. Liver CT sequence segmentation based with improved U-Net and graph cut. *Expert Syst. Appl.* **2019**, *126*, 54–63. [CrossRef]
13. Gao, Y.; Zhou, M.; Metaxas, D.N. UTNet: A Hybrid Transformer Architecture for Medical Image Segmentation. *arXiv* **2021**, arXiv:2107.00781.
14. Wang, Z.; Zou, Y.; Liu, P.X. Hybrid dilation and attention residual U-Net for medical image segmentation. *Comput. Biol. Med.* **2021**, *134*, 104449. [CrossRef] [PubMed]
15. Li, Y.; Wang, Z.; Yin, L.; Zhu, Z.; Qi, G.; Liu, Y. X-Net: A dual encoding–decoding method in medical image segmentation. *Vis. Comput.* **2021**, 1–11. [CrossRef]
16. Hatamizadeh, A.; Yang, D.; Roth, H.R.; Xu, D. UNETR: Transformers for 3D Medical Image Segmentation. *arXiv* **2021**, arXiv:2103.10504.
17. Cao, H.; Wang, Y.; Chen, J.; Jiang, D.; Zhang, X.; Tian, Q.; Wang, M. Swin-Unet: Unet-like Pure Transformer for Medical Image Segmentation. *arXiv* **2021**, arXiv:2105.05537.
18. Liu, Z.; Lin, Y.; Cao, Y.; Hu, H.; Wei, Y.; Zhang, Z.; Lin, S.; Guo, B. Swin transformer: Hierarchical vision transformer using shifted windows. *arXiv* **2021**, arXiv:2103.14030.
19. Girum, K.B.; Créhange, G.; Hussain, R.; Lalande, A. Fast interactive medical image segmentation with weakly supervised deep learning method. *Int. J. Comput. Assist. Radiol. Surg.* **2020**, *15*, 1437–1444. [CrossRef]
20. Wang, G.; Li, W.; Zuluaga, M.A.; Pratt, R.; Patel, P.A.; Aertsen, M.; Doel, T.; David, A.L.; Deprest, J.A.; Ourselin, S.; et al. Interactive Medical Image Segmentation Using Deep Learning with Image-Specific Fine Tuning. *IEEE Trans. Med. Imaging* **2018**, *37*, 1562–1573. [CrossRef]

21. Weng, G.; Dong, B.; Lei, Y. A level set method based on additive bias correction for image segmentation. *Expert Syst. Appl.* **2021**, *185*, 115633. [CrossRef]
22. Landman, B.; Xu, Z.; Igelsias, J.E.; Styner, M.; Langerak, T.; Klein, A. MICCAI multi-atlas labeling beyond the cranial vault–workshop and challenge. In Proceedings of the MICCAI: Multi-Atlas Labeling Beyond Cranial Vault-Workshop Challenge, Munich, Germany, 5–9 October 2015.
23. Kingma, D.P.; Ba, J. Adam: A Method for Stochastic Optimization. *CoRR* **2015**. CoRR:abs/1412.6980.
24. Gutmann, M.U.; Hyvärinen, A. *Noise-Contrastive Estimation: A New Estimation Principle for Unnormalized Statistical Models*; In Proceedings of the AISTATS, Sardinia, Italy, 13–15 May 2010.
25. van den Oord, A.; Li, Y.; Vinyals, O. Representation Learning with Contrastive Predictive Coding. *arXiv* **2018**, arXiv:1807.03748.
26. Paszke, A.; Gross, S.; Massa, F.; Lerer, A.; Bradbury, J.; Chanan, G.; Killeen, T.; Lin, Z.; Gimelshein, N.; Antiga, L.; et al. PyTorch: An Imperative Style, High-Performance Deep Learning Library; In Proceedings of the 2019 NeurIPS, Vancouver, BC, Canada, 8–14 December 2019.
27. Biewald, L. Experiment Tracking with Weights and Biases. 2020. Available online: wandb.com (accessed on 1 October 2021).

*applied
sciences*

Article

Advantages of Machine Learning in Forensic Psychiatric Research—Uncovering the Complexities of Aggressive Behavior in Schizophrenia

Lena A. Hofmann *, Steffen Lau and Johannes Kirchebner

Department of Forensic Psychiatry, University Hospital of Psychiatry, University of Zurich, Lenggstrasse, 8032 Zurich, Switzerland; steffen.lau@pukzh.ch (S.L.); Johannes.kirchebner@pukzh.ch (J.K.)
* Correspondence: lena.hofmann@pukzh.ch (L.H.); Tel.: +41-043-258-35-55

Abstract: Linear statistical methods may not be suited to the understanding of psychiatric phenomena such as aggression due to their complexity and multifactorial origins. Here, the application of machine learning (ML) algorithms offers the possibility of analyzing a large number of influencing factors and their interactions. This study aimed to explore inpatient aggression in offender patients with schizophrenia spectrum disorders (SSDs) using a suitable ML model on a dataset of 370 patients. With a balanced accuracy of 77.6% and an AUC of 0.87, support vector machines (SVM) outperformed all the other ML algorithms. Negative behavior toward other patients, the breaking of ward rules, the PANSS score at admission as well as poor impulse control and impulsivity emerged as the most predictive variables in distinguishing aggressive from non-aggressive patients. The present study serves as an example of the practical use of ML in forensic psychiatric research regarding the complex interplay between the factors contributing to aggressive behavior in SSD. Through its application, it could be shown that mental illness and the antisocial behavior associated with it outweighed other predictors. The fact that SSD is also highly associated with antisocial behavior emphasizes the importance of early detection and sufficient treatment.

Keywords: machine learning; advanced statistics; schizophrenia; aggression; forensic psychiatry

Citation: Hofmann, L.A.; Lau, S.; Kirchebner, J. Advantages of Machine Learning in Forensic Psychiatric Research—Uncovering the Complexities of Aggressive Behavior in Schizophrenia. *Appl. Sci.* **2022**, *12*, 819. https://doi.org/10.3390/app 12020819

Academic Editors: Kyungtae Kang, Hyo-Joong Suh and Junggab Son

Received: 19 November 2021
Accepted: 4 January 2022
Published: 14 January 2022

Publisher's Note: MDPI stays neutral with regard to jurisdictional claims in published maps and institutional affiliations.

1. Introduction

With the rapid technological progress of the past few years, artificial intelligence (AI) is increasingly being put to use in medical research. Often equated with human-like robots by the general public, AI is ultimately any system that adapts its performance based on its perception of the environment. This includes advanced statistics such as machine learning (ML), which allows a variety of variables and their relationship to one another to be analyzed through complex mathematical algorithms, as well as the quantification of the quality of a statistical model [1–4]. When it comes to psychiatric research though, statistical analyses are usually conducted using null hypothesis significance tests (NHSTs) or simple linear regressions. This results in the following certain limitations: (I) mainly linear relationships can be determined, and, with NHSTs, it is not even possible to examine the relationships between the variables themselves; (II) in order to avoid alpha error accumulation, only a limited number of variables can be analyzed; and (III) the research question must be precisely defined and rather constrained, as it can only be determined whether a (null) hypothesis is true or not. However, this approach does not accommodate psychiatric syndromes with their complex and often highly interdependent multifactorial relationships. The genesis of psychiatric diseases and pathological behavioral disorders is by no means a linear process influenced by only single, independent factors. This is especially true for the generally under-researched field of forensic psychiatry, where the interplay of psychopathology, offending, and aggression has yet to be comprehensively understood. Consequently, to investigate such phenomena, modern statistical methods such as ML are necessary and already applied in psychiatric research areas regarding pharmaceuticals or neuroimaging [5–9]. The following analysis should serve as an example of the practical use of ML in the field of forensic psychiatry, specifically aggression and schizophrenia spectrum disorders (SSDs). Factors linked to aggressive behavior outside the clinical setting have recently been evaluated by means of ML and include higher PANSS scores as well as younger age at SSD diagnosis [10–12]. For this study, we chose an explorative approach, as aggression is considered to be a multifactorial, complex phenomenon, mediated through a broad variety of parameters from different domains. This study now aims to determine the most predictive factors of aggression within the institutional setting, based on a unique group of forensic offenders with SSD, (objective I) and to quantify the performance of the calculated model (objective II).

2. Materials and Methods

The files of 370 delinquent patients diagnosed with SSD according to ICD-9 (295.x) [13] and ICD-10 (F20–29.x) [14], who were admitted to the Center for Inpatient Forensic Therapies of the University Hospital of Psychiatry Zurich, were assessed retrospectively. This comprehensive dataset included items from the following domains: social-demographic data, childhood/youth experiences, psychiatric history, past criminal history, social/sexual functioning, details on the offense leading to forensic hospitalization, prison data, and particularities of the current hospitalization and psychopathological symptoms. The latter was defined by an adapted positive and negative symptom scale (PANSS), whereby symptoms were divided into the usual 30 sub-categories and rated on a three-tier scale instead of a seven-tier one (completely absent, discretely present, or substantially present). The dataset has already been used in other studies as part of a larger, ongoing project with the goal of providing insights into the complex field of SSD and offending. Although the same database provides the basis for several analyses covering a wide range of objectives in this research area, and although there are a few overlapping parameters, it still contains a substantial number of unique variables, thus resulting in different theoretical and practical conclusions and implications. An overview of the basic characteristics of the population is provided in Table 1.

Table 1. Sociodemographic characteristics [1].

Characteristics	Total *n*/N (%)	No Aggression *n*/N (%)	Aggression *n*/N (%)
Male sex	327/352 (92.9)	219/239 (91.6)	108/113 (95.6)
Age at admission (mean, SD)	33.98 (10.206)	34.62 (10.014)	32.64 (10.519)
Native Country Switzerland	156/352 (44.3)	106/239 (44.4)	50/113 (44.2)
Single (at offense)	285/346 (82.4)	188/233 (80.7)	97/113 (85.8)

[1] SD = standard deviation; N = total study population; *n* = subgroup with characteristic.

Parts of the following section were published in advance in a study by Günther et al. [15] and is here partly replicated and extended by the methodology of the current research question. For further information regarding data collection and processing, please refer to previous publications [15–17]. Due to the explorative nature of this study, supervised machine learning (ML) seemed to be the optimal approach to identify the most relevant predictive factors out of numerous parameters and to determine the model providing the best predictive power. An overview of the statistical steps can be seen in Figure 1 and is further described below. All the steps were performed using R version 3.6.3. (R Project, Vienna, Austria) and the MLR package v2.171 (Bischl, Munich, Germany). CI calculations of the balanced accuracy were conducted using MATLAB R2019a (MATLAB and Statistics Toolbox Release 2012, The MathWorks, Inc., Natick, Massachusetts, United States) with the add-on "computing the posterior balanced accuracy" v1.0.

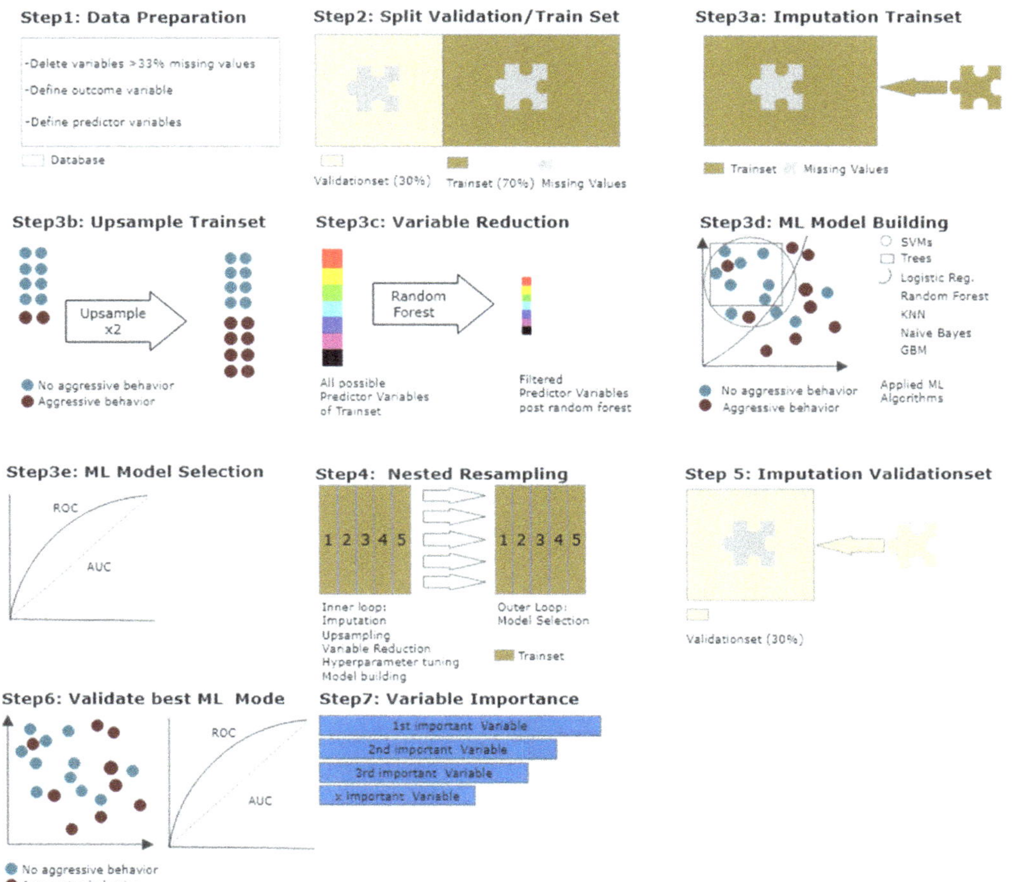

Figure 1. Overview of statistical procedures: (**Step 1**)—Data Preparation: Multiple categorical variables were converted to binary code. Continuous and ordinal variables were not manipulated. Outcome variable violent behavior/no violent behavior and 507 predictor variables were defined. (**Step 2**)—Data splitting: Split into 70% training dataset and 30% validation dataset. (**Step 3a–e**) —Model building and testing on training data I: Imputation by mean/mode; up-sampling of outcome "violent behavior" × 2; variable reduction via random forest; model building via ML algorithms —logistic regression, trees, random forest, gradient boosting, KNN (k-nearest neighbor), support vector machines (SVM) and naive Bayes; testing (selection) of best ML algorithm via ROC parameters. (**Step 4**)—Model building and testing on training data II: Nested resampling with imputation, up-sampling, variable reduction, and model building in inner loop and model testing on the outer loop. (**Step 5**)—Model building and testing on validation data I: Imputation with stored weights from Step 3a. (**Step 6**)—Model building and testing on validation data II: Best model identified in Step 3e applied on imputed and validation dataset and evaluated via ROC parameters. (**Step 7**)—Ranking of variables by indicative power.

All raw data were first processed for machine learning (see Figure 1, Step 1): Several categorical variables were converted to binary code, while continuous and ordinal variables were not adjusted. Due to the retrospective nature of the study and a large number of variables included, there were missing values among variables. This especially applied to information on the broader biographical history of patients, although forensic records were comprehensive. Variables with more than 33% missing values were eliminated, leaving a set of 508 variables. The outcome variable "aggressive behavior during current hospitalization" was dichotomized into (1) "aggressive behavior" and (0) "no aggressive behavior". Acts of aggression were defined as either verbal or physical attacks aimed toward staff or other patients, as well as damage of property. After the exclusion of all patients with missing information regarding their aggressive behavior from further analysis, a total of 352 patients remained. Out of these patients, 113 (32.1%) were involved in an aggressive event, while 239 (67.9%) were not (see Table 1). "No Aggression" was defined as the positive class, "Aggression" as the negative class.

After the completion of data preparation, the database was divided into one training and one validation subset (see Figure 1, Step 2). The training subset, including 70% of all cases ($n = 246$), was used for variable reduction and model building/selection. To enable the flexible application of all ML algorithms, imputation of missing values was carried out and imputation weights saved for later were reused on the validation subset (see Figure 1, Step 3a). As the outcome variable was unevenly distributed (12.4:87.6%), a random up-sampling at a rate of 2 was conducted, leading to a more balanced outcome (see Figure 1, Step 3b). A major objective of the present study was to identify the most important predictor variables from 507 possible variables. Additionally, a decrease in variables can counteract overfitting and maintain computing times in initial model building at an acceptable level. For this purpose, variable reduction to the 10 most important predictors was performed using randomForestSRC implemented in the MLR package (see Figure 1, Step 3c). As the database was relatively small for ML purposes and our focus lay on variable extraction and prediction, we applied discriminative model building with logistic regression, trees, random forest, gradient boosting, KNN (k-nearest neighbor), support vector machines (SVM), and as an easily applicable generative model building, naive Bayes (see Figure 1, Step 3d). No hyperparameters were optimized. The model performance of each model was calculated and assessed in terms of its balanced accuracy (the average of true positive and true negative rate, better suited for model evaluation and calculation of confidence intervals in imbalanced data) and goodness of fit (measured with the receiver operating characteristic, balanced curve area under the curve method, ROC balanced AUC). Specificity, sensitivity, positive predictive value (PPV), and negative predictive value (NPV) were also evaluated. As our training dataset was artificially balanced, the model with the highest AUC was chosen for final model validation with the test subset (see Figure 1, Step 3e). The set of identified variables was tested for multicollinearity to avoid dependencies between the variables. Finally, a nested resampling approach was employed, thus preventing the common obstacle of overfitting in ML. This was achieved using a nested resampling model with the inner loop performing imputation, oversampling, variable filtration, and model building within 5-fold cross-validation, and the outer loop for performance evaluation also embedded in 5-fold cross-validation—a technique for artificially creating different subsamples of a dataset (see Figure 1, Step 4).

As a next step, the validation subset, including 30% of all cases ($n = 106$), was applied to evaluate the statistical model selected before (see Figure 1, Steps 5–7). As briefly mentioned above, the previously stored imputation weights were reused on the validation subset (see Figure 1, Step 5). Then the model selected through the application of the training subset was applied for validation (see Figure 1, Step 6). The identified variables were finally tested for multicollinearity and ranked by their indicative power (see Figure 1, Step 7).

3. Results

3.1. Model Calculation

An overview of the performance parameters of the different calculated models during the nested resampling procedure can be found in Table 2. With a balanced accuracy of 77.6% and an AUC of 0.87, the support vector machines (SVM) outperformed all the other ML algorithms (see Table 2).

Table 2. Machine learning models and performance in nested cross-validation on training dataset [1].

Statistical Procedure	Balanced Accuracy (%)	AUC	Sensitivity (%)	Specificity (%)	PPV (%)	NPV (%)
Logistic Regression	74.9	0.85	77.8	72.1	85.7	60.6
Tree	74.7	0.80	72.6	76.8	86.2	56.7
Random Forest	75.3	0.83	74.9	74.9	87.3	59.9
Gradient Boosting						
KNN	77.7	0.85	78.6	76.8	88.0	63.1
SVM	**77.6**	**0.87**	**78.2**	**66.9**	**87.3**	**66.3**
Naive Bayes	75.9	0.85	87.9	76.1	87.8	59.8

[1] AUC = area under the curve (level of discrimination); PPV = positive predictive value; NPV = negative predictive value; KNN = k-nearest neighbors; SVM = support vector machines.

The absolute and relative distribution of the 10 most predictive variables identified during nested resampling and used for the model buildings can be seen in Table 3. They can be grouped in the following two areas: (1) problematic or antisocial behavior during current hospitalization, and (2) psychopathology. In the initial model including ten variables, the *time spent at a high-security level during current forensic hospitalization* was the dominant variable. However, the variable was omitted as it was considered circular. Further analysis was, therefore, conducted with the remaining nine most predictive variables.

Table 3. Absolute and relative distribution of relevant predictor variables [1].

Variable Code	Variable Description	Aggressive Incidents	No Aggressive Incidents
DZ1	Did the patient complain about the hospital staff?	73/111 (65.8)	45/238 (18.9)
DZ2	Did the patient show negative behavior toward other patients?	76/112 (67.9)	40/237 (16.9)
DZ7	Did the patient show dis/antisocial behavior?	90/111 (81.1)	73/238 (30.7)
DZ10	Did the patient break the rules of the ward (e.g., substance abuse)?	61/112 (54.5)	36/238 (15.1)
R22c (mean, SD)	Time spent at a high-security level during current forensic hospitalization	48.36 (59.65))	33.84 (45.22)
PA_A (mean, SD)	Adapted PANSS at admission: Total score	30.19 (12.34)	22.05 (11.35)
PA7	Adapted PANSS at admission: Hostility		
	symptom absent	27/113 (23.9)	160/238 (67.2)
	symptom discreet	22/113 (19.5)	45/238 (18.9)
	symptom substantial	64/113 (56.6)	33/238 (13.9)
PA18	Adapted PANSS at admission: Tension		
	symptom absent	25/113 (22.1)	131/238 (55)
	symptom discreet	25/113 (22.1)	54/238 (22.7)
	symptom substantial	63/113 (55.8)	53/238 (22.3)
PA22	Adapted PANSS at admission: Uncooperativeness		

Table 3. *Cont.*

Variable Code	Variable Description	Aggressive Incidents	No Aggressive Incidents
	symptom absent	22/113 (19.5)	144/238 (60.5)
	symptom discreet	38/113 (33.6)	58/238 (24.4)
	symptom substantial	53/113 (46.9)	36/238 (15.1)
PA28	Adapted PANSS at admission: Poor impulse control		
	symptom absent	17/113 (15)	155/238 (65.1)
	symptom discreet	33/113 (29.2)	40/238 (16.8)
	symptom substantial	63/113 (55.8)	43/238 (18.1)

[1] SD = Standard deviation; PANSS = positive and negative syndrome scale.

The quality of the final model in the validation step is shown in Table 4. As expected, the balanced accuracy of 73.5 and the AUC of 0.84 were less than the results of the initial training model but they were still meaningful. With a sensitivity of 84% and a specificity of 59%, the patients involved in aggressive incidents were identified correctly in almost three-quarters of events, while three-thirds of all the non-aggressive patients were identified correctly (see Table 4).

Table 4. Final SVM model performance measures on validation dataset.

Performance Measures	% (95% CI)
Balanced Accuracy	73.5 (64.4–82.1)
AUC	0.84 (0.75–0.93)
Sensitivity	83.5 (83.3–83.8)
Specificity	59.4 (58.8–59.9)
PPV	83.5 (83.2–83.8)
NPV	59.4 (58.8–59.9)

3.2. Determinants of Aggressive Inpatient Behavior

The distribution of the importance of variables of the final validation model is presented in Figure 2 as a one-sided tornado graph. *Negative behavior toward other patients* was identified as the most indicative factor in distinguishing aggressive and non-aggressive patients, followed by *breaking of ward rules*, the *PANSS score at admission*, and *poor impulse control* as well as *hostility*, also according to the PANSS. *Complaints about hospital staff, dis/antisocial utterances or attitudes, tension,* and *uncooperativeness* were also identified as factors influencing the model.

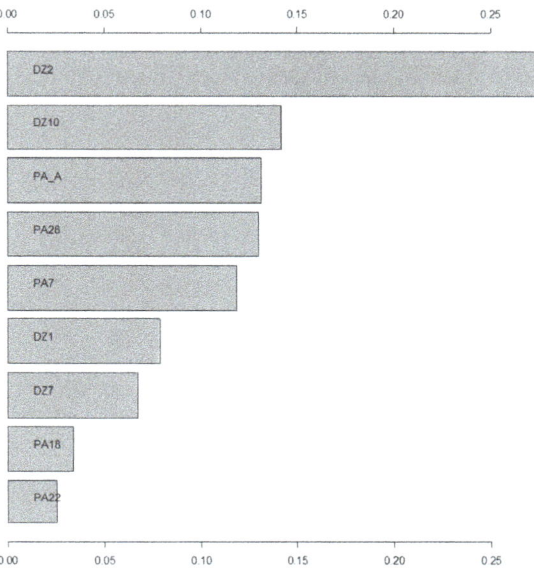

Figure 2. Importance of variable of final model "aggression" vs. "no aggression": DZ2 = patient showed negative behavior toward other patients; DZ10 = patient broke ward rules; PA_A = PANSS score at admission; PA28 = Adapted PANSS at admission: poor impulse control; PA7 = Adapted PANSS at admission: hostility; DZ1 = patient complained about the hospital staff; DZ7 = Patient showed dis/antisocial utterances or attitudes; PA18 = Adapted PANSS at admission: tension; PA22 = Adapted PANSS at admission: uncooperativeness.

4. Discussion

The purpose of this study was to identify the factors that distinguish between offender patients with SSD who show aggressive behavior within a hospitalized setting and those who do not. The idea was to exploratively identify the most predictive factors in inpatient violence. By applying ML algorithms to a large database, we were able to create an appropriate model consisting of nine factors. With a balanced accuracy of 73.5 and an AUC of 0.84, the model was able to correctly identify aggressive patients in nearly three-quarters of events and non-aggressive patients in two-thirds of events. The variables related to psychopathology and antisocial behavior proved to be the most predictive regarding inpatient aggression. The aggressive patients in our population showed an increased occurrence of *negative behavior toward other patients*. Patient/patient interaction is known to trigger aggressive events on general psychiatric wards in about a quarter of cases [16–18]. It seems obvious that this factor is all the more important in a forensic psychiatric hospital, where severely ill patients with a high potential for violence come together in a confined space with little opportunity for avoidance. This finding emphasizes the importance of de-escalating skills among staff. Remarkably, in contrast to previous results regarding acute general psychiatric wards, *negative behavior toward staff* was not identified as one of the ten most predictive parameters [19–22]. This seems somewhat contradictory to the fact that both *failure to comply with the ward rules* and *complains about hospital staff* were highly relevant in distinguishing aggressive from non-aggressive patients: Pointing out or insisting on adherence to ward rules and disciplining noncompliance often causes friction between staff and patients. While such situations arise on a regular basis within a highly institutionalized setting, such as forensic psychiatry, they do not seem to be obligatorily linked to the development of a negative attitude toward the staff and may be tolerated in the presence of a sustainable therapeutic relationship. *Poor impulse control, tension, hostility, and uncooperativeness*, measured using the corresponding PANSS scales, as well as the overall *PANSS score at admission*, were also identified as key factors related to inpatient aggression. While the results regarding a lack of impulse control, tension, and hostility are in line with previous findings regarding inpatient aggression in SSD patients as well as aggressive events prior to hospitalization, the link between overall symptomatology represented by the total PANSS score and aggression remains controversial [10,23–25]. This suggests that it is not the severity of disease alone that determines the development of aggression, but rather the interplay of the various factors present [23,25]. In summary, the factors that constitute aggression during hospitalization can be reduced to two domains, psychopathology, and antisocial behavior. Interestingly, these two domains outweighed all the other factors, for example, the parameters related to child development, social contacts, and family situation. This was surprising as childhood poverty, for instance, has been previously identified as a risk factor for violent offending [10,26,27]. This is similarly true for comorbid substance abuse, which has been identified as a risk factor for inpatient violence, especially in SSD patients, but did not prove to be of high influence in our population [11,26–29]. One possible explanation for these phenomena is that the highly structured and closely supervised setting of the forensic psychiatric institution compensates for social and biographical factors, as patients have little exposure to a potentially harmful original social environment (e.g., negative peer group, domestic conflicts, availability of drugs). In addition, it is worth considering that as SSD progresses and becomes more chronic, factors related to psychopathology may become more prominent than they are at an earlier stage of the disease and then outweigh factors with greater influence. As outlined above, in an initial analysis, the time spent at a high-security level during current hospitalization was identified as the most predictive factor outweighing the other variables by far. Since this was a circular argument, the item was omitted. Nevertheless, it should be noted that aggressive behavior in the context of hospitalization leads to a longer length of stay in high-security settings, which has both personal consequences for the patient regarding their rehabilitation and economic consequences for the healthcare system [30–32]. When interpreting these findings, the following two hypotheses deserve to be discussed: On the one hand, a conglomerate of SSD and antisocial traits might be present in the patients who display aggression in a highly institutionalized setting. The role of a potential comorbid antisocial personality disorder in SSD patients in the development of aggressive behavior has been extensively discussed in the literature [33–36]. On the other hand, antisociality may not be an expression of a comorbid personality disorder, but an expression of the underlying SSD. It is well known that positive psychotic symptoms such as hallucinations or threat/control–override symptoms can contribute to violent behavior [37–39]. This seems contradictory to the fact that the PANSS, regarding positive symptoms, was not identified as a risk factor in this study and further exploration is needed to distinguish between psychopathology and antisociality. Regarding the limitations of the present study, the most obvious lies in its retrospective design and, therefore, the small possibility of collecting selected parameters in a standardized manner. This poses a particular difficulty for the parameters that are hard to define, such as "antisocial behavior", since individual assessments of certain events may differ among different professionals. To minimize the possible bias effects and to draw robust causal inferences, a replication of the present findings in a prospective design is recommended. Furthermore, while a sample size of 370 patients is rather large regarding the field of forensic psychiatry, it has to be acknowledged that the dataset is a rather small one regarding medical research in other disciplines using ML algorithms. It is, therefore, recommended to apply the model to others. While the model was able to correctly identify aggressive patients in three-quarters of events, it failed to identify a third of all the non-aggressive patients as such. Therefore, one must be cautious when drawing definitive conclusions from the model, especially so if it is applied to clinical practice, where the label "aggressive patient" may affect a course of treatment, e.g., through a lower threshold for coercive measures. Thus, the question arises as to what value the AUC must assume in order to be considered an acutely acceptable performance measure. This discourse must be conducted intensively, particularly in the sensitive field of forensic psychiatry, since aggressive events occurring in such a setting have far-reaching consequences for the patients concerned (for example, compulsory medication, restrictions on freedom in the form of isolation and restraint as well as prolongation of hospitalization) [40].

5. Conclusions

Our findings expand the current research on factors influencing aggression within forensic inpatient treatment in offender patients with SSD. The present study is a good example of the practical use of artificial intelligence and illustrates that ML is instrumental in analyzing a large dataset and understanding the complex interplay between the factors that contribute to aggressive behavior in SSD. By applying ML, the 9 most predictive variables could be singled out from 507 items, and their interactions could be analyzed in an exploratory manner. A similar analysis with all 507 items would not have been feasible using linear regressions or even multivariate analyses, as the item number exceeds the capacities of those models, and the interplay of variables cannot be explored. In this study, we could show that mental illness and the antisocial behavior associated with it outweighed all the other factors. That these two groups have emerged as predictor domains is encouraging in that they are clinically elicitable using fairly simple methods. Biographical factors such as childhood trauma, on which psychiatrists are often focused when trying to explain aggressive behavior, are, in contrast, rather difficult to assess if the patients are not transparent and are also static, meaning they are perennially present regardless of the patients' individual development. Of course, these findings do not allow other known risk factors to be disregarded—yet they are outweighed by mental illness and antisocial behavior. The fact that SSD is also highly associated with antisocial behavior emphasizes the importance of early detection and sufficient treatment. The prevention of aggressive behavior toward fellow patients and staff members is a major concern in everyday clinical practice. Above all, but not exclusively, this applies to forensic psychiatric institutions in which a pre-selected group of patients with a particularly high risk of violence is treated. If the predictive risk domains are screened for, a tailor-made treatment approach could be designed for patients with an elevated risk. This may include closer monitoring by staff and case management by well-experienced therapists, who could even proactively develop skills counteracting aggressive impulses at an early stage with high-risk patients before the occurrence of such events.

Based on the present findings, the authors are currently developing a clinical screening tool for problematic inpatient behavior. Its application should enable clinicians to identify high-risk patients at an early stage, modify their treatment accordingly (for example, intensified monitoring), and ultimately prevent aggressive events during hospitalization. However, keeping the ethical implications described above in mind, one has to be mindful of the fact that despite this being a fairly large dataset in the niche subject of forensic psychiatry, these data can, for now, only serve as pilot data and need further application and exploration before a robust tool for detecting those patients with a high risk of aggressive behavior during hospitalization can be developed. In the future, this could not only protect staff and fellow patients from attacks, but also benefit the affected persons themselves, e.g., by reducing the need for coercive measures, shorter hospitalization, and the possible involvement of the judicial system.

Author Contributions: Conceptualization, J.K.; methodology, J.K.; software, J.K.; validation, J.K. and L.A.H.; formal analysis, J.K.; investigation, J.K. and L.A.H.; resources, L.A.H., J.K. and S.L.; data curation, J.K.; writing—original draft preparation, L.A.H.; writing—review and editing, L.A.H., S.L. and J.K.; visualization, L.A.H.; supervision, S.L. and J.K.; project administration, S.L. All authors have read and agreed to the published version of the manuscript.

Funding: This research received no external funding.

Institutional Review Board Statement: This study was reviewed and approved by the Ethics Committee Zurich [Kanton Zürich] (committee's reference number: KEK-ZH-NR 2014–0480). The study complied with the Helsinki Declaration of 1975, revised in 2008.

Informed Consent Statement: Patient consent was waived due to the retrospective design, for which formal consent is not required.

Data Availability Statement: The dataset generated and analyzed during the current study are available from the corresponding author on reasonable request.

Conflicts of Interest: The authors declare no conflict of interest.

References

1. Iniesta, R.; Stahl, D.; McGuffin, P. Machine learning, statistical learning and the future of biological research in psychiatry. *Psychol. Med.* **2016**, *46*, 2455–2465. [CrossRef] [PubMed]
2. Oquendo, M.A.; Baca-García, E.; Artés, A.; PerezCruz, F.; Galfalvy, H.; Blascofontecilla, H.; Madigan, D.; Duan, N. Machine learning and data mining: Strategies for hypothesis generation. *Mol. Psychiatry* **2012**, *17*, 956–959. [CrossRef]
3. Rutledge, R.B.; Chekroud, A.M.; Huys, Q.J. Machine learning and big data in psychiatry: Toward clinical applications. *Curr. Opin. Neurobiol.* **2019**, *55*, 152–159. [CrossRef] [PubMed]
4. Stephan, K.E.; Mathys, C. Computational approaches to psychiatry. *Curr. Opin. Neurobiol.* **2014**, *25*, 85–92. [CrossRef] [PubMed]
5. Chekroud, A.M.; Bondar, J.; Delgadillo, J.; Doherty, G.; Wasil, A.; Fokkema, M.; Cohen, Z.; Belgrave, D.; DeRubeis, R.; Iniesta, R.; et al. The promise of machine learning in predicting treatment outcomes in psychiatry. *World Psychiatry* **2021**, *20*, 154–170. [CrossRef] [PubMed]
6. Janssen, R.J.; Mourão-Miranda, J.; Schnack, H.G. Making Individual Prognoses in Psychiatry Using Neuroimaging and Machine Learning. *Biol. Psychiatry Cogn. Neurosci. Neuroimaging* **2018**, *3*, 798–808. [CrossRef] [PubMed]
7. Dwyer, D.B.; Falkai, P.; Koutsouleris, N. Machine Learning Approaches for Clinical Psychology and Psychiatry. *Annu. Rev. Clin. Psychol.* **2018**, *14*, 91–118. [CrossRef] [PubMed]
8. Gillan, C.; Whelan, R. What big data can do for treatment in psychiatry. *Curr. Opin. Behav. Sci.* **2017**, *18*, 34–42. [CrossRef]
9. Bzdok, D.; Meyer-Lindenberg, A. Machine Learning for Precision Psychiatry: Opportunities and Challenges. *Biol. Psychiatry: Cogn. Neurosci. Neuroimaging* **2017**, *3*, 223–230. [CrossRef] [PubMed]
10. Sonnweber, M.; Lau, S.; Kirchebner, J. Violent and non-violent offending in patients with schizophrenia: Exploring influences and differences via machine learning. *Compr. Psychiatry* **2021**, *107*, 152238. [CrossRef] [PubMed]
11. Fazel, S.; Långström, N.; Hjern, A.; Grann, M.; Lichtenstein, P. Schizophrenia, substance abuse, and violent crime. *JAMA* **2009**, *301*, 2016–2023. [CrossRef] [PubMed]
12. Fazel, S.; Smith, E.N.; Chang, Z.; Geddes, J.R. Risk factors for interpersonal violence: An umbrella review of meta-analyses. *Br. J. Psychiatry* **2018**, *213*, 609–614. [CrossRef]
13. WHO. *ICD-9: International Classification of Diseases*; 9th Revision; World Health Organization: Geneva, Switzerland, 1978.
14. WHO. *ICD-10: International Statistical Classification of Diseases and Related Health Problems: Tenth Revision*, 5th ed.; World Health Organization: Geneva, Switzerland, 2016.
15. Kirchebner, J.; Lau, S.; Kling, S.; Sonnweber, M.; Günther, M.P. Individuals with schizophrenia who act violently towards others profit unequally from inpatient treatment-Identifying subgroups by latent class analysis. *Int. J. Methods Psychiatr. Res.* **2020**, *30*, e1856. [CrossRef] [PubMed]
16. Papadopoulos, C.; Ross, J.; Stewart, D.; Dack, C.; James, K.; Bowers, L. The antecedents of violence and aggression within psychiatric in-patient settings. *Acta Psychiatr. Scand.* **2012**, *125*, 425–439. [CrossRef]
17. Edwards, J.G.; Jones, D.; Reid, W.H.; Chu, C.C. Physical assaults in a psychiatric unit of a general hospital. *Am. J. Psychiatry* **1988**, *145*, 1568–1571.
18. Chou, K.-R.; Lu, R.-B.; Chang, M. Assaultive Behavior by Psychiatric In-Patients and Its Related Factors. *J. Nurs. Res.* **2001**, *9*, 139–151. [CrossRef] [PubMed]
19. Nijman, H.L.; Allertz, W.F.; Merckelbach, H.L.; Ravelli, D.P. Aggressive behaviour on an acute psychiatric admissions ward. *Eur. J. Psychiatry* **1997**, *11*, 106–114.
20. Omérov, M.; Wistedt, B. Mangageable violence in a new ward for acutely admitted patients. *Eur. Psychiatry* **1997**, *12*, 311–315. [CrossRef]
21. Mellesdal, L. Aggression on a Psychiatric Acute Ward: A Three-Year Prospective Study. *Psychol. Rep.* **2003**, *92* (Suppl. S3), 1229–1248. [CrossRef] [PubMed]
22. Omerov, M.; Edman, G.; Wistedt, B. Incidents of violence in psychiatric inpatient care. *Nord. J. Psychiatry* **2002**, *56*, 207–213. [CrossRef] [PubMed]
23. Nolan, K.A.; Volavka, J.; Czobor, P.; Sheitman, B.; Lindenmayer, J.-P.; Citrome, L.L.; McEvoy, J.; Lieberman, J.A. Aggression and psychopathology in treatment-resistant inpatients with schizophrenia and schizoaffective disorder. *J. Psychiatr. Res.* **2005**, *39*, 109–115. [CrossRef]

24. Arango, C.; Barba, A.C.; González-Salvador, T.; Ordóñez, A.C. Violence in Inpatients With Schizophrenia: A Prospective Study. *Schizophr. Bull.* **1999**, *25*, 493–503. [CrossRef] [PubMed]
25. Fresán, A.; Apiquian, R.; de la Fuente-Sandoval, C.; Löyzaga, C.; García-Anaya, M.; Meyenberg, N.; Nicolini, H. Violent behavior in schizophrenic patients: Relationship with clinical symptoms. *Aggress. Behav.* **2005**, *31*, 511–520. [CrossRef]
26. Camus, D.; Gholam, M.; Conus, P.; Bonsack, C.; Gasser, J.; Moulin, V. Individual and contextual factors associated with violent behaviours during psychiatric hospitalizations. *Encephale* **2021**. [CrossRef]
27. Camus, D.; Glauser, E.S.D.; Gholamrezaee, M.; Gasser, J.; Moulin, V. Factors associated with repetitive violent behavior of psychiatric inpatients. *Psychiatry Res.* **2021**, *296*, 113643. [CrossRef] [PubMed]
28. Hodgins, S.; Tiihonen, J.; Ross, D. The consequences of Conduct Disorder for males who develop schizophrenia: Associations with criminality, aggressive behavior, substance use, and psychiatric services. *Schizophr. Res.* **2005**, *78*, 323–335. [CrossRef]
29. Whiting, D.; Lichtenstein, P.; Fazel, S. Violence and mental disorders: A structured review of associations by individual diagnoses, risk factors, and risk assessment. *Lancet Psychiatry* **2021**, *8*, 150–161. [CrossRef]
30. Kirchebner, J.; Günther, M.P.; Sonnweber, M.; King, A.; Lau, S. Factors and predictors of length of stay in offenders diagnosed with schizophrenia—A machine-learning-based approach. *BMC Psychiatry* **2020**, *20*, 201. [CrossRef] [PubMed]
31. Davoren, M.; Byrne, O.; O'Connell, P.; O'Neill, H.; O'Reilly, K.; Kennedy, H.G. Factors affecting length of stay in forensic hospital setting: Need for therapeutic security and course of admission. *BMC Psychiatry* **2015**, *15*, 301. [CrossRef]
32. de Tribolet-Hardy, F.; Habermeyer, E. Schizophrenic Patients between General and Forensic Psychiatry. *Front. Public Health* **2016**, *4*, 135. [CrossRef] [PubMed]
33. Hodgins, S.; Müller-Isberner, R. Schizophrenie und Gewalt. *Nervenarzt* **2014**, *85*, 273–278. [CrossRef] [PubMed]
34. Hodgins, S. Violent behaviour among people with schizophrenia: A framework for investigations of causes, and effective treatment, and prevention. *Philos. Trans. R. Soc. B Biol. Sci.* **2008**, *363*, 2505–2518. [CrossRef]
35. Lau, S.; Günther, M.; Kling, S.; Kirchebner, J. Latent class analysis identified phenotypes in individuals with schizophrenia spectrum disorder who engage in aggressive behaviour towards others. *Eur. Psychiatry* **2019**, *60*, 86–96. [CrossRef] [PubMed]
36. Hodgins, S.; Toupin, J.; Côté, G. Schizophrenia and antisocial personality disorder: A criminal combination. In *Explorations in Criminal Psychopathology: Clinical Syndromes with Forensic Implications*; Charles C Thomas Publisher: Springfield, IL, USA, 1996.
37. Swanson, J.W.; Van Dorn, R.A.; Swartz, M.S.; Smith, A.; Elbogen, E.B.; Monahan, J. Alternative pathways to violence in persons with schizophrenia: The role of childhood antisocial behavior problems. *Law Hum. Behav.* **2008**, *32*, 228–240. [CrossRef] [PubMed]
38. Chan, B.; Shehtman, M. Clinical risk factors of acute severe or fatal violence among forensic mental health patients. *Psychiatry Res.* **2019**, *275*, 20–26. [CrossRef]
39. Bo, S.; Abu-Akel, A.; Kongerslev, M.; Haahr, U.H.; Simonsen, E. Risk factors for violence among patients with schizophrenia. *Clin. Psychol. Rev.* **2011**, *31*, 711–726. [CrossRef]
40. Hotzy, F.; Theodoridou, A.; Hoff, P.; Schneeberger, A.R.; Seifritz, E.; Olbrich, S.; Jäger, M. Machine Learning: An Approach in Identifying Risk Factors for Coercion Compared to Binary Logistic Regression. *Front. Psychiatry* **2018**, *9*, 258. [CrossRef] [PubMed]

Article

Automated Extraction of Cerebral Infarction Region in Head MR Image Using Pseudo Cerebral Infarction Image by CycleGAN

Mizuki Yoshida [1], Atsushi Teramoto [1,*], Kohei Kudo [2], Shoji Matsumoto [3], Kuniaki Saito [1] and Hiroshi Fujita [4]

[1] Graduate School of Health Sciences, Fujita Health University, Toyoake 470-1192, Japan; mint.9511@gmail.com (M.Y.); saitok@fujita-hu.ac.jp (K.S.)
[2] Daido Hospital, Nagoya 457-8511, Japan; k.kudo1206@gmail.com
[3] Faculty of Medicine, Fujita Health University, Toyoake 470-1192, Japan; shoji.neuro@gmail.com
[4] Faculty of Engineering, Gifu University, Gifu 501-1194, Japan; fujita@fjt.info.gifu-u.ac.jp
* Correspondence: teramoto@fujita-hu.ac.jp

Abstract: Since recognizing the location and extent of infarction is essential for diagnosis and treatment, many methods using deep learning have been reported. Generally, deep learning requires a large amount of training data. To overcome this problem, we generated pseudo patient images using CycleGAN, which performed image transformation without paired images. Then, we aimed to improve the extraction accuracy by using the generated images for the extraction of cerebral infarction regions. First, we used CycleGAN for data augmentation. Pseudo-cerebral infarction images were generated from healthy images using CycleGAN. Finally, U-Net was used to segment the cerebral infarction region using CycleGAN-generated images. Regarding the extraction accuracy, the Dice index was 0.553 for U-Net with CycleGAN, which was an improvement over U-Net without CycleGAN. Furthermore, the number of false positives per case was 3.75 for U-Net without CycleGAN and 1.23 for U-Net with CycleGAN, respectively. The number of false positives was reduced by approximately 67% by introducing the CycleGAN-generated images to training cases. These results indicate that utilizing CycleGAN-generated images was effective and facilitated the accurate extraction of the infarcted regions while maintaining the detection rate.

Keywords: cerebral infarction; CycleGAN; deep learning

Citation: Yoshida, M.; Teramoto, A.; Kudo, K.; Matsumoto, S.; Saito, K.; Fujita, H. Automated Extraction of Cerebral Infarction Region in Head MR Image Using Pseudo Cerebral Infarction Image by CycleGAN. *Appl. Sci.* **2022**, *12*, 489. https://doi.org/10.3390/app12010489

Academic Editors: Manuel Armada, Kyungtae Kang, Hyo-Joong Suh and Junggab Son

Received: 18 November 2021
Accepted: 2 January 2022
Published: 4 January 2022

Publisher's Note: MDPI stays neutral with regard to jurisdictional claims in published maps and institutional affiliations.

1. Introduction

Stroke is a leading cause of death globally [1]. Cerebral infarction, the most common type of stroke [2], often has after-effects and affects the quality of life. Early detection and treatment are essential for cerebral infarction because the infarcted region expands over time.

Computed tomography (CT) and magnetic resonance imaging (MRI) are mainly used to diagnose cerebral infarction. MRI is widely used today because of its high contrast resolution and ability to visualize brain structures and lesions. MR modalities, such as T2WI, FLAIR, and diffusion-weighted imaging (DWI), are mainly used to diagnose cerebral infarction. DWI is excellent for detecting a stroke during the hyperacute phase due to the high signal of reduced diffusion caused by cellular edema [3]. Therefore, the detection of acute cerebral infarction by DWI is useful for prompt diagnosis and for treatment selection. However, DWI has a lower resolution than other sequences, and its imaging principle tends to cause artifacts and distortion, making it challenging to identify the presence or absence, as well as the extent, of infarction. Moreover, stroke specialists are not always present during emergencies, and it may be too complex to make an accurate diagnosis.

Here, we focus on computer-aided diagnosis (CAD). CAD uses image processing to detect and analyze lesions and is used by physicians for second opinions on diagnoses. The beginning of CAD was mainly based on image analysis and machine learning [4]. Later, the concept of deep learning was proposed [5], and its application fields have been broadened to include speech recognition, natural language processing, and image recognition [6]. In the field of image recognition, it has been actively used for classification, regression, and segmentation tasks [7]. In particular, segmentation has become an important task in the medical field, to recognize organs and to understand the location and extent of lesions. FCN [8], SegNet [9], and U-Net [10] are commonly used deep learning techniques for segmentation. In particular, U-Net has shown good performance in segmenting medical images, and has been applied to many regions [11–13].

Since recognizing the location and extent of infarction is essential for diagnosis and treatment, several studies have been reported on the automatic detection and segmentation of cerebral infarction [14–20]. Rajini et al. [14] proposed a method for detecting cerebral infarction in CT images, and Barros et al. [15] proposed a method for segmenting the infarcted cerebral region using a convolutional neural network (CNN). In addition, Chen et al. [16] segmented the cerebral infarct region from a DWI using a CNN. Dolz et al. [17] proposed a method for segmenting infarcted regions with U-Net using multiple image sequences. Paing et al. [19] recently proposed automated segmentation of the infarcted region using variational mode decomposition and U-Net. Segmentation performance was evaluated using a total of 239 cases from a public dataset, and the results showed high similarity to the gold standard. Furthermore, Zhang et al. [20] proposed stroke lesion detection using a deep learning model for object detection.

Generally, deep learning requires a large amount of training data. However, the collection of medical imaging data is sometimes limited by ethical and other issues. Data augmentation is often used to prevent overfitting because of the small amount of data. During data augmentation, the number of images is increased by image manipulations, such as rotation, enlargement, contraction, contrast change, and the addition of noise. However, data augmentation does not significantly change the nature of the lesion or the target structure, and it is not expected to increase the amount of data.

To overcome this problem, we focussed on generative adversarial networks (GANs) [21]. GAN is a network model that generates images similar to training image data and was proposed by Goodfellow et al. in 2014. Recently, deep convolutional GAN [22], information maximizing GAN [23], Wasserstein GAN [24], and CycleGAN [25] have been developed as derivative technologies for GANs. Among them, CycleGAN can transform images without paired training data; it can convert MR images to CT images [26] and reduce noise [27]. Recently, several studies on the mutual conversion of images using CycleGAN have been performed. However, to the best of our knowledge, only a few studies have reported using CycleGAN-generated images for the extensive training of deep learning models. If CycleGAN generates brain images from normal brain MR images, it is expected to improve the accuracy of extraction and segmentation by increasing the variety of data.

In this study, we aimed to develop a novel method for virtually generating cerebral infarction images from healthy images using CycleGAN and applying them to the U-Net training model to improve the cerebral infarction automatic extraction accuracy by U-Net.

Hereafter, U-Net with CycleGAN represents an approach that uses CycleGAN-generated images for training; U-Net without CycleGAN represents an approach that uses only real images.

2. Materials and Methods

2.1. Outline

Figure 1 shows an outline of the proposed method. Normal brain MR images were converted into cerebral infarction images using CycleGAN. Subsequently, U-Net is trained using the CycleGAN-generated images and images of actual patients with cerebral infarction, and the infarcted regions were automatically extracted.

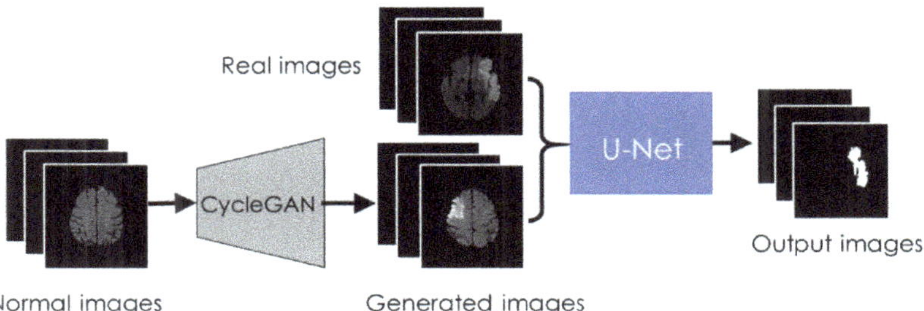

Figure 1. Study outline schematic.

2.2. Image Dataset

This study collected DWI of MR images acquired at the Fujita Health University Hospital and Daido Hospital, including 64 healthy cases (1280 images) and 160 cerebral infarction cases (4788 images).

Figure 2 shows the distribution of the cases. The mean area of the infarcted region was 922.54 ± 1169.25 mm^2. The mean signal intensity was 91.78 ± 14.75 within the healthy region and 161.70 ± 24.81 within the infarcted region. Figure 2a shows the distribution of the differences between the signal intensities of the healthy and infarcted regions. The difference between the signal intensities was calculated by subtracting the mean of the pixel intensities within the infarcted and healthy regions. The mean difference between the signal intensities of the infarcted and healthy regions was 69.91 ± 22.76.

As ground truth, a binary image was created. The pixel intensities of the infarcted regions were set at 255, whereas the background was set at 0. The infarcted region was also confirmed by a radiological technologist with more than 10 years of clinical experience. As a basic data augmentation technique, the number of images was doubled by the left-right flipping operation in this study. Examples of the collected images and ground truth are shown in Figure 3.

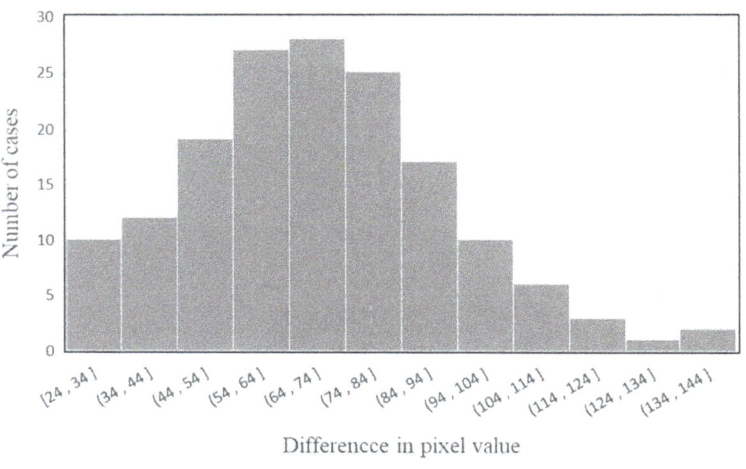

(**a**)

Figure 2. *Cont.*

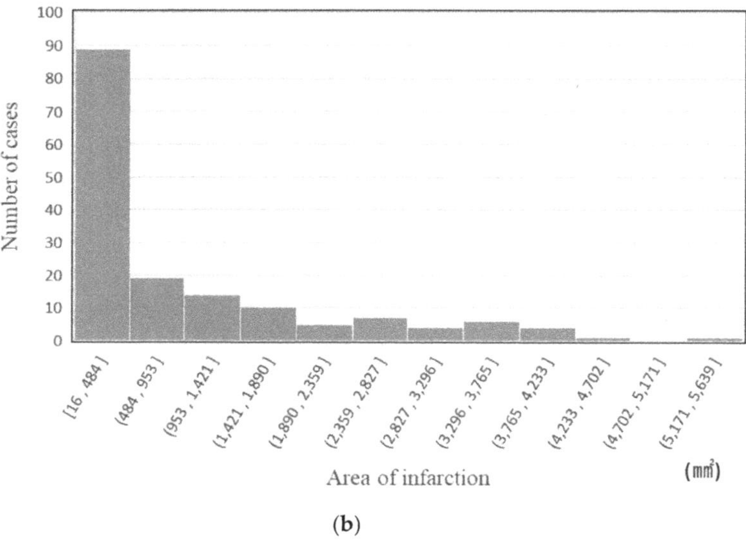

(**b**)

Figure 2. Distribution of cases. (**a**) Distribution by difference in pixel values; (**b**) Distribution by area of infarction.

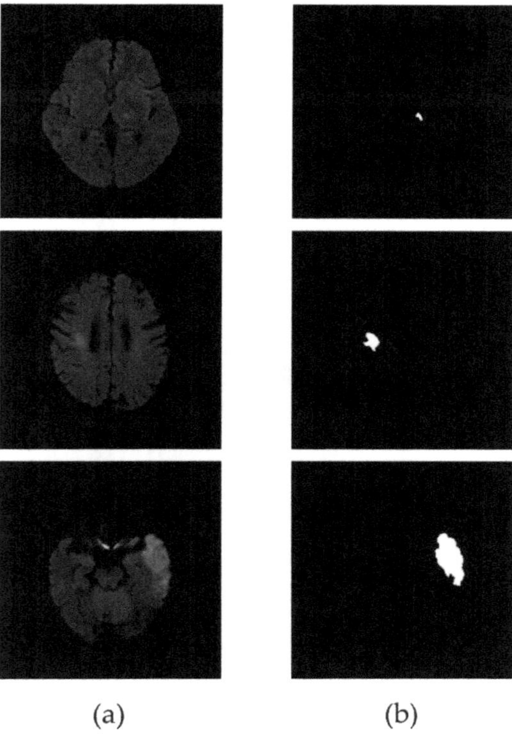

(a)　　　　　　(b)

Figure 3. Example of original images and ground truths. (**a**) Original image; (**b**) Ground truth.

2.3. Generation of Pseudo Abnormal Images by CycleGAN

We generated a pseudo-patient image from a healthy image using domain transformation techniques. Domain translation techniques include Pix2Pix and CycleGAN, etc. Pix2Pix requires paired images for training, but it is difficult to obtain paired images of a healthy image and a diseased image. Therefore, we used CycleGAN, which can perform image translation without paired images. The structure of CycleGAN is shown in Figure 4.

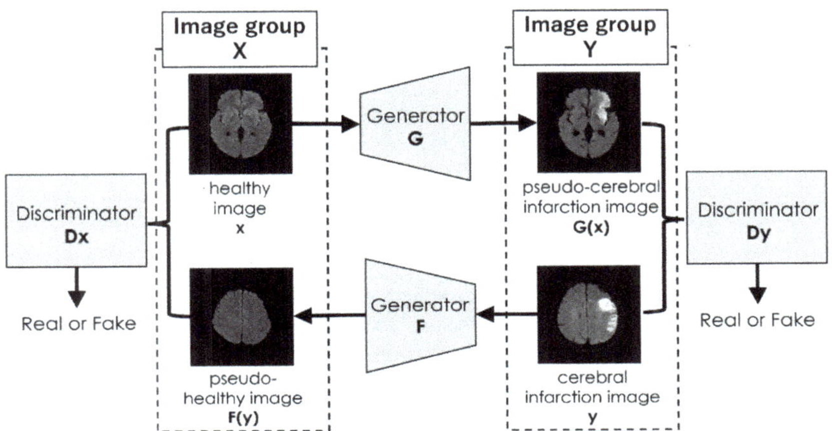

Figure 4. CycleGAN structure.

The CycleGAN structure used in this study followed the algorithm proposed by Zhu et al. [25]. CycleGAN consists of two generators and two discriminators. The two generators convert one image group to another. The discriminator determines whether the data transformed by the generator and the actual data are real or fake. Once an image has been detected as real or fake, the generator trains to perform a transformation that will result in images as close to real as possible. CycleGAN uses cycle consistency loss, in addition to the adversarial loss used in normal GANs. The cycle consistency loss was calculated by comparing the distributions generated by the cycle based on the training data.

We prepared two image groups: 243 slices from 52 stroke patients with infarction and 300 slices in 15 of 64 healthy cases taken at the Daido Hospital. CycleGAN was trained to convert images of the healthy and infarction cases. After training, the stroke pseudo-images were generated following the provision of healthy images to the generator of the CycleGAN, which converts images of healthy cases to those of infarction cases. Furthermore, we created a ground truth on the generated pseudo infarction images. First, the infarcted region was identified by subtracting the image before conversion from the generated image. Then, the noise generated by the subtraction was manually removed. In addition, pixel value of the infarct area was set to 255 and the background was set to 0 in a binary image, as in the correct image of the real image.

A nine-block ResNet [28] was used for the CycleGAN generator, and the structure of PatchGAN [29] was used as the discriminator. To train the CycleGAN, we implemented an original python program using TensorFlow and Keras that are the deep learning libraries. The number of epochs was set to 200. The learning rate was set to 0.0002, and the batch size was set to 1. The training and testing were executed on a computer equipped with a graphical processing unit (NVIDIA GeForce GTX TITAN X).

2.4. Extraction of Infarcted Region

To extract the infarcted region, U-Net, a deep learning model, was used for segmentation. In this study, we chose U-Net as a preliminary study because it had achieved many results in segmentation. The U-Net structure used in this study is shown in Figure 5. U-Net is a semantic segmentation method that was presented at the Medical Image Computing and Computer-Assisted Intervention in 2015. This network achieved good results during the cell segmentation challenge of the International Symposium on Biomedical Imaging in 2015. This network is an extension of the Fully Convolutional Network and allows accurate segmentation with less training data. The U-Net structure is shown in Figure 5, and it is called U-Net because of its U-shaped network structure. The left half of the U-shape is called the encoder, and the right half is called the decoder. The encoder was composed of a convolutional and pooling layer. In the former layer, convolutional operations were performed to extract features from the given image. Subsequently, the output was passed through a rectified linear unit, an activation function. In the latter layer, max-pooling was used to downscale the image features. Thus, the features were extracted and compressed through several convolutional and pooling layers. In the decoder, they were upsampled by deconvolution. The high-resolution features obtained from the encoder were combined with the up-sampled output, and a convolutional operation was performed. Furthermore, the decoder concatenated and cropped the output of the encoder at the same depth. This process ensured that information was propagated from the encoder to the decoder at all scales, and no information was lost during the down-sampling operations of the encoder. The input was an image with a matrix size of height (H) = 256 and width (W) = 256; the infarcted region had pixel intensity values >0. The training was performed on 100 cases of cerebral infarction at the Daido Hospital and Fujita Medical University Hospital. Fifty test cases were randomly selected from the Fujita Health University Hospital cases that were not used for CycleGAN and U-Net training.

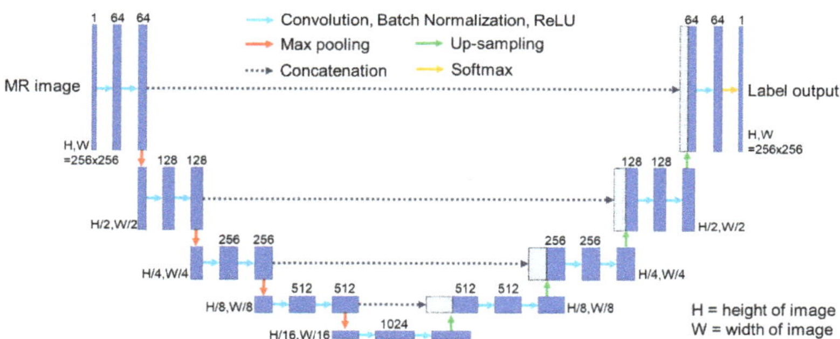

Figure 5. U-Net structure.

In this study, each convolutional layer of the encoder and decoder had five layers. A batch normalization layer was added at the end of each layer. We developed an original python program using TensorFlow and Keras for the training of the U-Net. The number of epochs was set to 300, with a training rate of 0.0001 and a batch size of 32. The training and testing were executed on a computer equipped with a graphical processing unit (NVIDIA TITAN RTX).

2.5. Evaluation Metrics

The detection and segmentation accuracies were evaluated to verify the effectiveness of the proposed method. First, we defined the criteria for evaluation as follows.

The infarcted region was detected if there was an overlap between the infarcted region extracted by U-Net and the infarcted region in the ground truth. This evaluation was conducted in 3D.

The number of false positives (FPs) was calculated using healthy images. We counted the number of extracted regions divided by the number of cases to obtain the number of FPs per case. The number of extracted regions was automatically calculated using 3D labeling.

DI and Jaccard index (JI) were employed to evaluate the extraction accuracy. DI and JI were used to assess the similarity between the extracted region and the ground truth (ideal infarction region), and they were calculated using the following equations:

$$Dice\ index(A, B) = \frac{2|A \cap B|}{|A| + |B|} \tag{1}$$

$$Jaccard\ index(A, B) = \frac{|A \cap B|}{|A \cup B|} \tag{2}$$

where A indicates the ground truth region and B indicates the extracted region of the cerebral infarction.

3. Results

3.1. CycleGAN-Generated Images

Examples of the CycleGAN-generated cerebral infarction images are shown in Figure 6. Figure 6a shows a healthy image before conversion by CycleGAN, Figure 6b shows a pseudo cerebral infarction image using CycleGAN. Images with large infarct regions are shown in case 1 and 2, and those with small infarct regions are shown in Figure 6 case 3. Figure 6 case 4 shows an example of failure in which an infarct was not generated, and the pixel intensity values of the entire image were high.

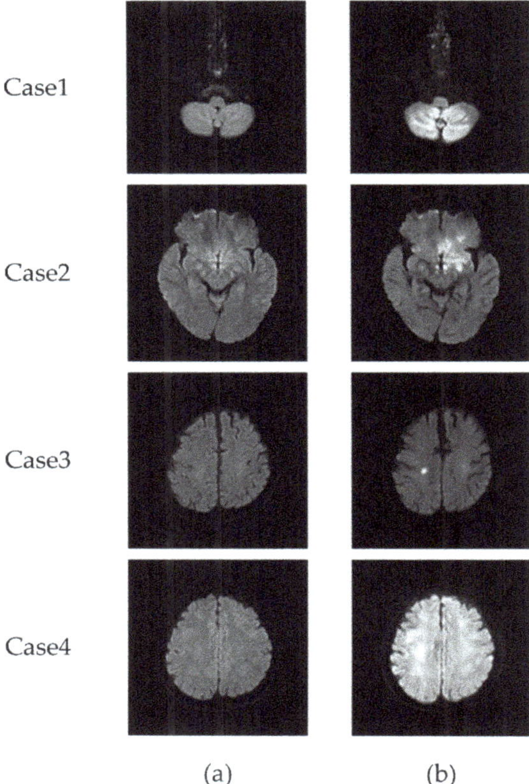

Figure 6. Examples of cerebral infarction CycleGAN-generated images. (**a**) Healthy image; (**b**) Pseudo cerebral infarction image. (The generated images are shown at each brain height. (Case 1) medulla oblongata level; (Case 2) midbrain level; (Cases 3 and 4) cortical level.).

3.2. Extraction of the Infarcted Regions

The results of the U-Net extraction of the infarcted region are shown in Figure 7. The extraction accuracy and sensitivity of cerebral infarction are shown in Table 1. Regarding extraction accuracy, the U-Net Dice index (DI) was 0.473 without CycleGAN and 0.553 with CycleGAN, showing an improvement of approximately 8%. Figure 8 compares the extraction accuracies for different amounts of training cases. In the cases of 25 to 100 trainings, the extraction accuracy was higher when the CycleGAN-generated images were added to the training cases.

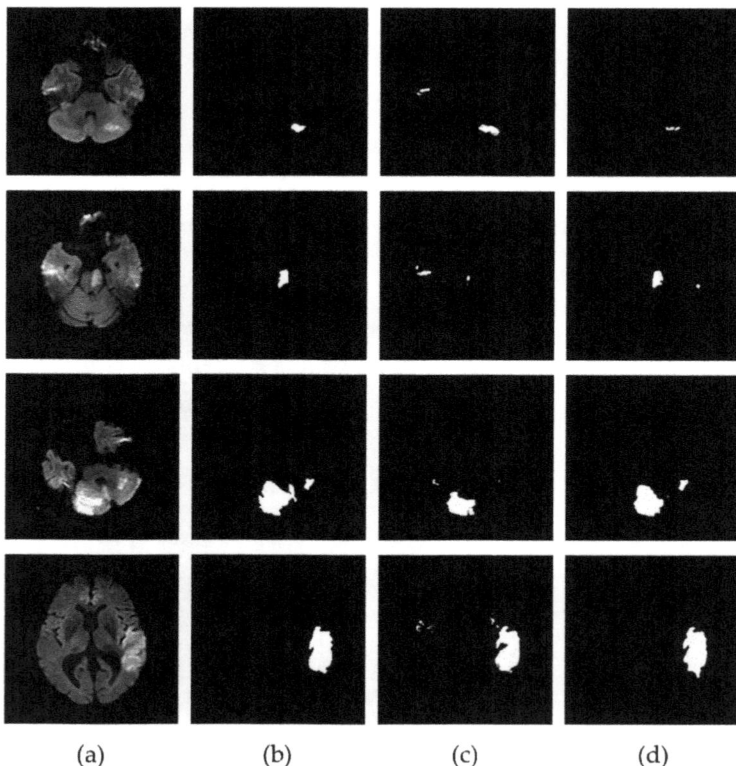

(a)   (b)   (c)   (d)

Figure 7. Extraction outcomes of the infarcted region. (**a**) Original image; (**b**) Ground truth; (**c**) U-Net without CycleGAN; (**d**) U-Net with CycleGAN.

Table 1. Extraction accuracy and sensitivity of cerebral infarction.

	U-Net without CycleGAN	**U-Net with CycleGAN**
Dice index	0.473	0.553
Jaccard index	0.360	0.433
Sensitivity	0.940	0.920
False positives per case	3.750	1.234

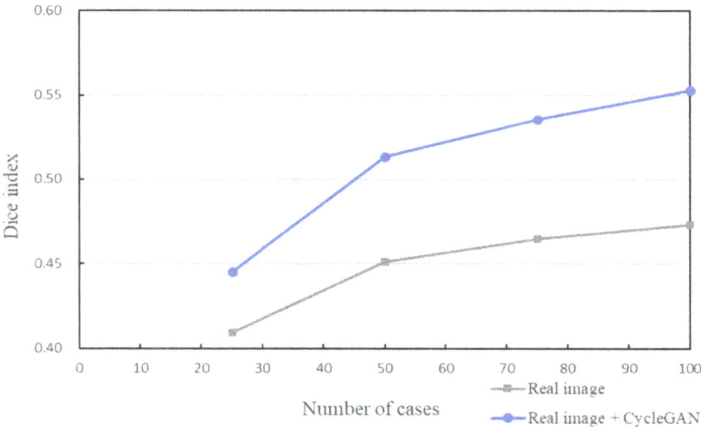

Figure 8. Comparison of Dice index for different numbers of study cases.

The U-Net cerebral infarction sensitivity was 0.94 without CycleGAN and 0.92 with CycleGAN. The detection results using 64 healthy cases with no infarcted regions are shown in Figure 9. When only real images were used for training, the number of FPs per case was 3.75. On the other hand, it was 1.23 when CycleGAN-generated images were introduced. The number of FPs was reduced by approximately 67% by introducing CycleGAN-generated images to the training cases.

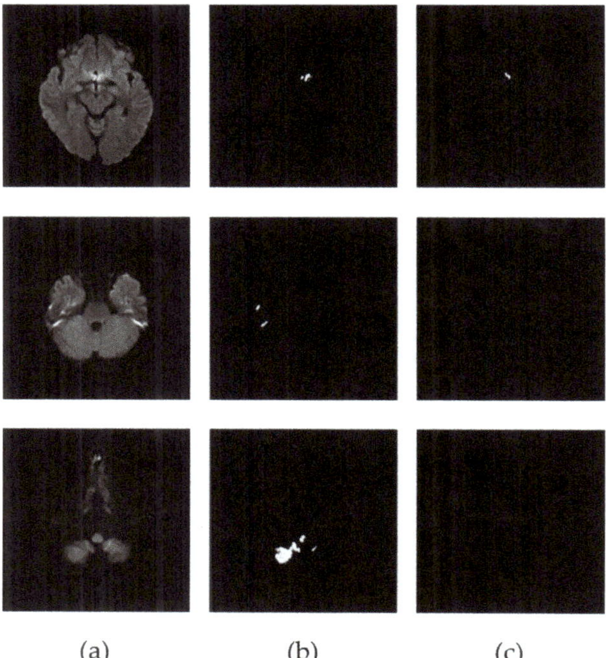

(a)   (b)   (c)

Figure 9. Extraction outcomes of healthy cases. (**a**) Original image; (**b**) U-Net without CycleGAN; (**c**) U-Net with CycleGAN.

4. Discussion

The CycleGAN-generated images ranged from small infarcts, such as lacunar infarcts, to large infarcts, such as cardiogenic cerebral emboli. Infarct images were also generated from brain slices at different heights; infarcts along the lateral ventricles and cerebellum were also generated. Most of the generated infarct images had high signal intensities within the infarcted region, and infarcts with low signal intensity were not often generated. The signal intensities of the infarcts used in the study tended to be high. In some of the generated images, the signal intensities of the entire image increased, and no infarct was generated.

We extracted the regions of cerebral infarction using the U-Net. Then, we compared the accuracy and sensitivity of extracting infarcted regions when only actual cases of stroke patients were used for U-Net training and when CycleGAN-generated images and actual cases were used. The DI for extraction accuracy was 0.473 and 0.553 when using only real cases and CycleGAN-generated images, respectively. The extraction accuracy improved when using generated images for training. One of the reasons for the improvement in the extraction accuracy is that the infarcted region characteristics extracted by U-Net changed when the CycleGAN-generated images were used together. Most of the CycleGAN-generated images had small and high-contrast infarcted regions. U-Net was trained using these images, and the infarcted regions with a difference of more than 80 from the healthy region were extracted, and the DI was 0.707. However, if the infarcted region was large and the signal value within the infarcted region was uneven, only the high-signal region was extracted and underestimated in some cases. The sensitivity for detecting infarcts was 0.940 when training on real cases only and 0.920 when using CycleGAN-generated images. The combined use of CycleGAN-generated images decreased the sensitivity because one small infarct was missed that had been detected when only real cases were used. The sensitivity differed depending on the size of the infarcted area and the difference in pixel values between the normal parenchyma and the infarcted area, both in real cases alone and when CycleGAN-generated images were used. The sensitivity of infarcts with diameters >10 mm and a pixel difference >50 was 0.976. However, infarcts with sizes of <10 mm or faint infarcts with pixel intensity differences of <50 were not extracted in some cases, and the sensitivity was 0.750. It is challenging to recognize contrast with the surrounding normal parenchyma in cases of small or low-contrast infarcts. Furthermore, when the number of FPs was compared, the number of FPs per case was 3.750 when only real cases were used and 1.234 when CycleGAN-generated images were used, a reduction of approximately 67% in the number of FPs. FPs were caused by a slight signal increase in the brain parenchyma and linear high-signal artifacts in the peripheral areas of the brain parenchyma due to DWI distortion. In many cases, a region with a relatively high signal compared to the surrounding area in the healthy brain parenchyma was mistakenly identified as the infarcted area. When the CycleGAN-generated images were combined, a high-signal area in the brain parenchyma was rarely identified as the wrong infarcted area. The CycleGAN introduced in this study generated many patterns that resembled actual infarcts. Therefore, data could be collected to more clearly separate the characteristics of FPs and infarct patterns, and U-Net, trained on infarct patterns generated by CycleGAN and actual infarcts, could distinguish between infarcts and FPs accurately. Although it is necessary to improve the detection of small and low-contrast infarcts, including those detected when training was performed only in real cases, the combined use of CycleGAN-generated images is highly effective in removing FPs. This method can be said to be effective overall.

Studies using deep learning require several images for training, and studies using large datasets, multi-sequence images, and 3D networks have been conducted to extract cerebral infarctions [16,17,30]. The accuracy comparison with the other groups is shown in Table 2. Among these, one study using a 3D network has a DI of nearly 0.8, which is inferior to the 3D network in terms of extraction accuracy. Compared with Chen et al. [16], who used DWI in the same 2D network, the extraction accuracy was lower, but the sensitivity was the same, and the number of FPs was lower in our method. By performing region extraction using only DWI, which is almost always performed at any institution for acute stroke diagnosis, the collection of cases is facilitated, and detection is not prevented due to missing data. Some studies have been performed to improve CycleGAN accuracy [31]. However, no studies have generated pseudo-patient lesion images. The results of our study suggest that performance can be improved even when only limited patient data are obtained. The method used in this study can be applied to rare diseases, with only a few reported cases.

Table 2. Comparison of accuracy with other groups.

	Sensitivity	DI	FP
RF (Mitra et al. [32])		0.53	
FCM (Muda et al. [33])		0.73	0.16
CNN (Chen et al. [16])	0.94	0.67	3.27
U-Net (Dolz et al. [17])		0.635	
U-Net (Paing et al. [19])		0.668	
U-Net without CycleGAN	0.94	0.473	3.75
U-Net with CycleGAN	0.92	0.553	1.23

Regarding the limitations of this method, cases from two facilities were used. Still, the images used for CycleGAN training were not used for the segmentation test data using U-Net, so the U-Net test cases were cases obtained at a single facility only. In the future, it will be necessary to increase the number of cases and collect data from multiple facilities to verify the results. In addition, subjective and quantitative evaluation are considered essential because we did not evaluate the detailed image quality of the generated images in this study. Furthermore, in this study, U-Net and CycleGAN were introduced as a preliminary study. In the future, it is necessary to use improved models of U-Net and CycleGAN as well as other models to improve the accuracy.

5. Conclusions

We developed a method to extract infarcted regions from head MR images using U-Net. Furthermore, the training images were augmented using CycleGAN. The results showed that the use of CycleGAN-generated images was effective for accurately extracting the infarcted region while maintaining the detection rate.

Author Contributions: Conceptualization: M.Y. and A.T.; methodology: M.Y. and A.T.; software: M.Y. and A.T.; validation: M.Y., K.K. and S.M.; formal analysis: M.Y.; investigation: M.Y. and A.T.; resources: A.T.; data curation: K.K. and S.M.; writing—original draft preparation: M.Y.; writing—review and editing: A.T., K.S. and H.F.; visualization: M.Y.; supervision: A.T.; project administration: A.T. and K.S. All authors have read and agreed to the published version of the manuscript.

Funding: This research received no external funding.

Institutional Review Board Statement: The study was conducted in accordance with the Declaration of Helsinki, and approved by the Institutional Review Board of Fujita Health University (HM20-489, 16 March 2021).

Informed Consent Statement: Informed consent was obtained via an opt-out process at the Fujita Health University Hospital and Daido Hospital, and all data were anonymized.

Data Availability Statement: The source code used to support the findings of this study will be available from the corresponding author upon request.

Acknowledgments: We are grateful to Ayumi Yamada and Yuya Onishi of Fujita Health University for helpful advice for this study.

Conflicts of Interest: The authors declare no conflict of interest.

References

1. Benjamin, E.J.; Blaha, M.J.; Chiuve, S.E.; Cushman, M.; Das, S.R.; Deo, R.; de Ferranti, S.D.; Floyd, J.; Fornage, M.; Gillespie, C.; et al. Heart disease and stroke statistics-2016 update a report from the American Heart Association. *Circulation* **2017**, *135*, e38–e48. [CrossRef] [PubMed]
2. Feigin, V.L.; Lawes, C.M.; Bennett, D.A.; Anderson, C. Stroke epidemiology: A review of population-based studies of incidence, prevalence, and case-fatality in the late 20th century. *Lancet Neurol.* **2003**, *2*, 43–53. [CrossRef]
3. Lutsep, H.L.; Albers, G.W.; Decrespigny, A.; Kamat, G.N.; Marks, M.P.; Moseley, M.E. Clinical utility of diffusion-weighted magnetic resonance imaging in the assessment of ischemic stroke. *Ann. Neurol.* **1997**, *41*, 574–580. [CrossRef] [PubMed]
4. Doi, K. Computer-aided diagnosis in medical imaging: Historical review, current status and future potential. *Comput. Med. Imaging Graph.* **2007**, *31*, 198–211. [CrossRef]
5. Hinton, G.E.; Salakhutdinovr, R.R. Reducing the Dimensionality of Data with Neural Networks. *Science* **2006**, *313*, 504–507. [CrossRef]
6. LeCun, Y.; Bengio, Y.; Geoffrey Hinton, G. Deep learning. *Nature* **2015**, *521*, 436–444. [CrossRef]
7. Hesamian, M.H.; Jia, W.; He, X.; Kennedy, P. Deep Learning Techniques for Medical Image Segmentation: Achievements and Challenges. *J. Digit. Imaging* **2019**, *32*, 582–596. [CrossRef]
8. Long, J.; Shelhamer, E.; Darrell, T. Fully convolutional networks for semantic segmentation. In Proceedings of the IEEE Conference on Computer Vision and Pattern Recognition, Boston, MA, USA, 7–12 June 2015; pp. 3431–3440. [CrossRef]
9. Badrinarayanan, V.; Kendall, A.; Cipolla, R. Segnet: A deep convolutional encoder-decoder architecture for image segmentation. *IEEE Trans. Pattern Anal. Mach. Intell.* **2017**, *39*, 2481–2495. [CrossRef] [PubMed]
10. Ronneberger, O.; Fischer, P.; Brox, T. U-Net: Convolutional Networks for Biomedical Image Segmentation. In Proceedings of the Medical Image Computing and Computer-Assisted Intervention–MICCAI 2015, Munich, Germany, 5–9 October 2015; pp. 234–241.
11. Dong, H.; Yang, G.; Liu, F.; Mo, Y.; Guo, Y. Automatic Brain Tumor Detection and Segmentation Using U-Net Based Fully Convolutional Networks. In *Communications in Computer and Information Science*; Springer: Cham, Switzerland, 2017; pp. 506–517. [CrossRef]
12. Seo, H.; Huang, C.; Bassenne, M.; Xiao, R.; Xing, L. Modified U-Net (mU-Net) with Incorporation of Object-Dependent High Level Features for Improved Liver and Liver-Tumor Segmentation in CT Images. *IEEE Trans. Med. Imaging* **2020**, *39*, 1316–1325. [CrossRef]
13. Gaál, G.; Maga, B.; Lukács, A. Attention U-Net Based Adversarial Architectures for Chest X-ray Lung Segmentation. In Proceedings of the Workshop on Applied Deep Generative Networks Co-Located with 24th European Conference on Artificial Intelligence 2020, CEUR Workshop Proceedings 2692, Santiago de Compostela, Spain, 29 August–8 September 2020.
14. Rajini, N.H.; Bhavani, R. Computer aided detection of ischemic stroke using segmentation and texture features. *Measurement* **2013**, *46*, 1865–1874. [CrossRef]
15. Barros, R.S.; Tolhuisen, M.; Boers, A.M.; Jansen, I.; Ponomareva, E.; Dippel, D.W.J.; Van Der Lugt, A.; Van Oostenbrugge, R.J.; Van Zwam, W.H.; Berkhemer, O.A.; et al. Automatic segmentation of cerebral infarcts in follow-up computed tomography images with convolutional neural networks. *J. NeuroInt. Surg.* **2019**, *12*, 848–852. [CrossRef]
16. Chen, L.; Bentley, P.; Rueckert, D. Fully automatic acute ischemic lesion segmentation in DWI using convolutional neural networks. *NeuroImage Clin.* **2017**, *15*, 633–643. [CrossRef] [PubMed]
17. Dolz, J.; Ben Ayed, I.; Desrosiers, C. Dense Multi-path U-Net for Ischemic Stroke Lesion Segmentation in Multiple Image Modalities. *Lect. Notes Comput. Sci.* **2019**, *11383*, 271–282. [CrossRef]
18. Karthik, R.; Menaka, R.; Johnson, A.; Anand, S. Neuroimaging and deep learning for brain stroke detection—A review of recent advancements and future prospects. *Comput. Methods Programs Biomed.* **2020**, *197*, 105728. [CrossRef]
19. Paing, M.; Tungjitkusolmun, S.; Bui, T.; Visitsattapongse, S.; Pintavirooj, C. Automated Segmentation of Infarct Lesions in T1-Weighted MRI Scans Using Variational Mode Decomposition and Deep Learning. *Sensors* **2021**, *21*, 1952. [CrossRef] [PubMed]
20. Zhang, S.; Xu, S.; Tan, L.; Wang, H.; Meng, J. Stroke Lesion Detection and Analysis in MRI Images Based on Deep Learning. *J. Health Eng.* **2021**, *2021*, 5524769. [CrossRef]
21. Goodfellow, I.; Pouget-Abadie, J.; Mirza, M.; Xu, B.; Warde-Farley, D.; Ozair, S.; Courville, A.; Bengio, Y. Generative adversarial nets. *Adv. Neural. Inf. Process Syst.* **2014**, *27*, 2672–2680.
22. Radford, A.; Metz, L.; Chintala, S. Unsupervised representation learning with deep convolutional generative adversarial networks. In Proceedings of the 4th International Conference on Learning Representations, ICLR 2016—Conference Track Proceedings, San Juan, PR, USA, 2–4 May 2016.
23. Chen, X.; Duan, Y.; Houthooft, R.; Schulman, J.; Sutskever, I.; Abbeel, P. InfoGAN: Interpretable representation learning by information maximizing generative adversarial nets. *Adv. Neural Inf. Process. Syst.* **2016**, *29*, 2180–2188.
24. Arjovsky, M.; Chintala, S.; Bottou, L. Wasserstein GAN. In Proceedings of the 34th International Conference on Machine Learning, Sydney, Australia, 6–11 August 2017; Volume 70, pp. 214–223.
25. Zhu, J.-Y.; Park, T.; Isola, P.; Efros, A.A. Unpaired image-to-image translation using cycle-consistent adversarial networks. In Proceedings of the IEEE International Conference on Computer Vision, Venice, Italy, 22–29 October 2017; pp. 2223–2232.
26. Hiasa, Y.; Otake, Y.; Takao, M.; Matsuoka, T.; Takashima, K.; Carass, A.; Prince, J.L.; Sugano, N.; Sato, Y. Cross-Modality Image Synthesis from Unpaired Data Using CycleGAN. *Adv. Data Min. Appl.* **2018**, *11037*, 31–41. [CrossRef]

27. Zhou, L.; Schaefferkoetter, J.D.; Tham, I.W.; Huang, G.; Yan, J. Supervised learning with cyclegan for low-dose FDG PET image denoising. *Med. Image Anal.* **2020**, *65*, 101770. [CrossRef]

28. He, K.; Zhang, X.; Ren, S.; Sun, J. Deep Residual Learning for Image Recognition. In Proceedings of the IEEE Conference on Computer Vision and Pattern Recognition, Las Vegas, NV, USA, 27–30 June 2016; pp. 770–778. [CrossRef]

29. Isola, P.; Zhu, J.-Y.; Zhou, T.; Efros, A.A. Image-to-Image Translation with Conditional Adversarial Networks. In Proceedings of the 2017 IEEE Conference on Computer Vision and Pattern Recognition (CVPR), Honolulu, HI, USA, 21–26 July 2017; pp. 5967–5976. [CrossRef]

30. Zhang, R.; Zhao, L.; Lou, W.; Abrigo, J.; Mok, V.C.T.; Chu, W.C.W.; Wang, D.; Shi, L. Automatic Segmentation of Acute Ischemic Stroke From DWI Using 3-D Fully Convolutional DenseNets. *IEEE Trans. Med. Imaging* **2018**, *37*, 2149–2160. [CrossRef]

31. Sandfort, V.; Yan, K.; Pickhardt, P.J.; Summers, R.M. Data augmentation using generative adversarial networks (CycleGAN) to improve generalizability in CT segmentation tasks. *Sci. Rep.* **2019**, *9*, 16884. [CrossRef] [PubMed]

32. Mitra, J.; Bourgeat, P.; Fripp, J.; Ghose, S.; Rose, S.; Salvado, O.; Connelly, A.; Campbell, B.; Palmer, S.; Sharma, G.; et al. Lesion segmentation from multimodal MRI using random forest following ischemic stroke. *NeuroImage* **2014**, *98*, 324–335. [CrossRef] [PubMed]

33. Muda, A.F.; Saad, N.; Bakar, S.; Muda, S.; Abdullah, A. Brain lesion segmentation using fuzzy C-means on diffusion-weighted imaging. *ARPN J. Eng. Appl. Sci.* **2015**, *10*, 1138–1144.

Article

STHarDNet: Swin Transformer with HarDNet for MRI Segmentation

Yeonghyeon Gu [†], Zhegao Piao [†] and Seong Joon Yoo *

Department of Computer Science and Engineering, Sejong University, Seoul 05006, Korea;
yhgu@sejong.ac.kr (Y.G.); piaozhegao5@gmail.com (Z.P.)
* Correspondence: sjyoo@sejong.ac.kr
† These authors contributed equally to this work.

Abstract: In magnetic resonance imaging (MRI) segmentation, conventional approaches utilize U-Net models with encoder–decoder structures, segmentation models using vision transformers, or models that combine a vision transformer with an encoder–decoder model structure. However, conventional models have large sizes and slow computation speed and, in vision transformer models, the computation amount sharply increases with the image size. To overcome these problems, this paper proposes a model that combines Swin transformer blocks and a lightweight U-Net type model that has an HarDNet blocks-based encoder–decoder structure. To maintain the features of the hierarchical transformer and shifted-windows approach of the Swin transformer model, the Swin transformer is used in the first skip connection layer of the encoder instead of in the encoder–decoder bottleneck. The proposed model, called STHarDNet, was evaluated by separating the anatomical tracings of lesions after stroke (ATLAS) dataset, which comprises 229 T1-weighted MRI images, into training and validation datasets. It achieved Dice, IoU, precision, and recall values of 0.5547, 0.4185, 0.6764, and 0.5286, respectively, which are better than those of the state-of-the-art models U-Net, SegNet, PSPNet, FCHarDNet, TransHarDNet, Swin Transformer, Swin UNet, X-Net, and D-UNet. Thus, STHarDNet improves the accuracy and speed of MRI image-based stroke diagnosis.

Keywords: ATLAS; HarDNet; Swin transformer; segmentation; U-Net

Citation: Gu, Y.; Piao, Z.; Yoo, S.J. STHarDNet: Swin Transformer with HarDNet for MRI Segmentation. *Appl. Sci.* **2022**, *12*, 468. https://doi.org/10.3390/app12010468

Academic Editors: Kyungtae Kang, Junggab Son and Hyo-Joong Suh

Received: 19 November 2021
Accepted: 1 January 2022
Published: 4 January 2022

Publisher's Note: MDPI stays neutral with regard to jurisdictional claims in published maps and institutional affiliations.

1. Introduction

Strokes pose a threat to human health because of their high incidence, mortality rate, and potential for causing disabilities. Strokes can be diagnosed using a variety of advanced testing methods, among which brain computed tomography (CT) or magnetic resonance imaging (MRI) are often used. The CT scan is the best method for classifying acute cerebral infarction and brain hemorrhage and is performed first in patients suspected of having strokes to determine initial treatment. In the case of cerebral infarction, it is displayed as a low density, and in the case of a stroke, it is displayed as high density. However, the infarction part does not appear well in the early stage of cerebral infarction. An MRI test is similar to a CT scan, but it has the advantage that it can accurately find small lesions or lesions in the brain region that are difficult to find in CT scans because it has much better imaging power. Many studies have been conducted on stroke diagnosis using computer vision technology to help doctors with diagnosis [1–3]. Conventional CT or MRI image-based diagnoses have often used U-Net models [4] with encoder–decoder structure using the convolution neural network (CNN) structure and obtained good results.

Recently, with the application of transformers to the computer vision field, many segmentation models using transformers have also been proposed [5,6]. A transformer is a successful example of applying the method of processing sequence data in natural language processing (NLP) analysis to the field of computer vision. Transformers currently exhibit good performance in the field of computer vision, including detection [7], segmentation [8], and classification [9]. Furthermore, models that have an encoder–decoder structure by combining a CNN model and a vision transformer (ViT) model have also performed well in medical image analysis [10,11]. Currently, many combination models of CNN and ViT use transformers in the bottleneck of the CNN-based encoder–decoder model so that parameters of the encoder and decoder can be delivered more effectively [12,13]. However, CNN-based encoder–decoder models are large in size and slow in terms of calculation speed, and ViT models have the problem that the calculation amount of the model increases sharply as the image size increases. One of the reasons why the transformer is used in the bottleneck in conventional combination models of CNN and ViT is to minimize the effect that the transformer has on the overall calculation speed as the size of the input feature map increases.

This study combines a HarDNet block [14], a lightweight model structure among CNN models, and a Swin Transformer model, which solves the problem of the calculation amount increasing sharply in the transformer model as the image size increases in the ViT model. This is to solve the existing problem and, at the same time, obtain high performance.

To this end, an encoder–decoder model in the form of U-Net is constructed with HarDNet blocks. Unlike a method that uses a transformer layer in the bottleneck of the encoder–decoder in past studies, this study uses a Swin Transformer [15] model in the first skip connection layer of the encoder model. As a result, the Swin Transformer model maintains the advantages of shifted windows approach based on self-attention and hierarchical feature extraction because a larger feature map is applied compared to when it is used in the bottleneck. The STHarDNet model proposed in this study has the following characteristics:

- Using HarDNet blocks, the proposed model improves slow computational speed, a disadvantage of the conventional CNN-based encoder–decoder structure model.
- Using the Swin Transformer, the proposed model solves the problem in which the memory use and computations increase as the image size increases in the ViT model.
- Using the first skip connection layer in the encoder of the Swin Transformer model in the encoder–decoder model that has the U-Net structure, it accepts a feature map larger than the bottleneck as input and maintains the Swin Transformer model's shifted windows approach and hierarchical transformer characteristics.

This study used the Anatomical Tracings of Lesions After Stroke (ATLAS) to conduct comparative experiments of performance. The ATLAS dataset is a standardized open dataset built for performance comparison of various algorithms that manually segment lesion locations in the MRI images of 229 stroke patients. In the ATLAS data, the MRI images of 177 patients were used as training data, and the data of the remaining 52 patients were used as validation data to conduct the comparative experiments of performance with existing state-of-the-art (SOTA) models. In the results, the proposed model's Dice, IoU, precision, and recall were 0.5547, 0.4185, 0.6764, and 0.5286, respectively, indicating that the proposed model performed better than the conventional model. Furthermore, it showed faster performance in the comparative calculation speed test of the models.

2. Related Work

2.1. HarDNet Block

The HarDNet Block consists of multiple harmonic dense blocks (HDBs). In the HarD-Net Block, a depthwise-separable convolution layer is used for connection between HDBs. This reduced the convolutional input/output (CIO) by 50% compared to when a 1×1 convolutional layer and a 2×2 average pooling layer were used in DenseNet [14]. In the connection method between HDBs, when the value of $k - 2^n$ is larger than 0, and the value of $\frac{2^n}{k}$ is a natural number, the k-th layer is connected to the $k - 2^n$-th layer, as shown in

Equation (1). In Equation (1), k is the position of the layer in HDB, n is the layer connected to k in the HDB, and N is a natural number.

$$C_k = k - 2^n, \ \ if \ \frac{2^n}{k} \in N, \ k - 2^n \geq 0 \tag{1}$$

2.2. Swin Transformer Block

The Swin Transformer consists of Swin transformer blocks, which are created to be suitable for detection and segmentation by introducing the concept of a hierarchical feature map and shifted windows to ViT. In conventional transformers, self-attention is performed by creating tokens with the same patch size, but the Swin Transformer uses a method of merging adjacent patches gradually, starting from a patch size of 4×4, like the hierarchical structure of the feature pyramid network. This allows using each hierarchical feature map's information, like U-Net. Figure 1 shows the patch merging process, where, in the figure, a red box refers to a window, the small box (1–16) refers to a patch (token) with a size of 4×4, and patch merging merges 2×2 patches into one. In the process of patch merging, the feature map's size is down-sampled to (W/2, H/2).

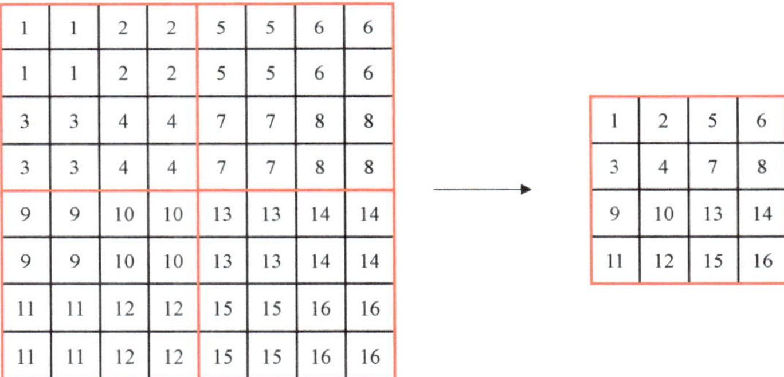

Figure 1. Example of patch merging in windows of the Swin Transformer.

For Swin transformer blocks, self-attention is performed in respective windows only, and merging is performed in the last feature map. This solves the problem of increasing computations when the image size increases in conventional ViT. However, the positions of the windows are fixed, and the relationship between the windows is not represented because self-attention is performed in the fixed windows only. Therefore, to calculate the relationship between two windows, the window is shifted to the right ($\rightarrow$) and down ($\downarrow$) directions by window size/2, and the self-attention is performed once more [15]. As a result, the Swin Transformer facilitates the analysis of the entire input image with self-attention alone in respective windows.

Figure 2 shows an example of the shifted windows approach. In Figure 2, a red box means a window, and a small box (1–16) means a patch (token) with a size of 4×4. In the figure, there are four windows with a window size of 4. In Swin transformer block 2, the windows are shifted based on Swin transformer block 1. In Swin transformer block 2, the windows are shifted in the right ($\rightarrow$) and down ($\downarrow$) directions by the window size/2, and the feature parts that the windows lack are supplemented, as shown in colors in the figure.

Figure 3 below shows a schematic diagram for connecting two Swin transformer blocks. As shown in the figure, the standard windows-based, multi-head, self-attention (W-MSA) module and shifted window-based, multi-head, self-attention (SW-MSA) module are used sequentially in the Swin transformer block. There is LayerNorm (LN) in front and back of S(W)-MSA, and the last MPL consists of two GELU non-linearities. Therefore, the Swin transformer block is used in multiples of two [16].

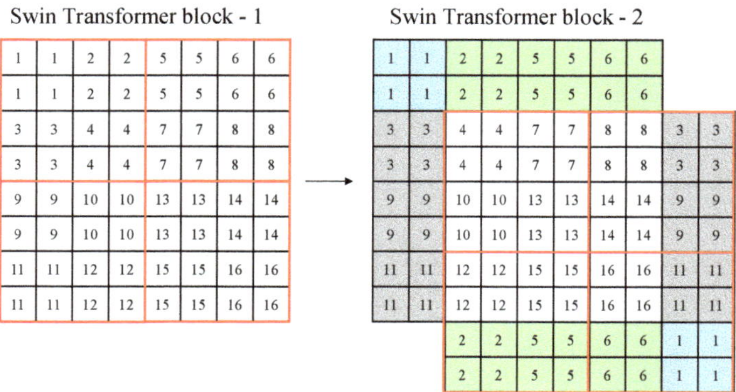

Figure 2. Example of a shifted window approach of the Swin Transformer.

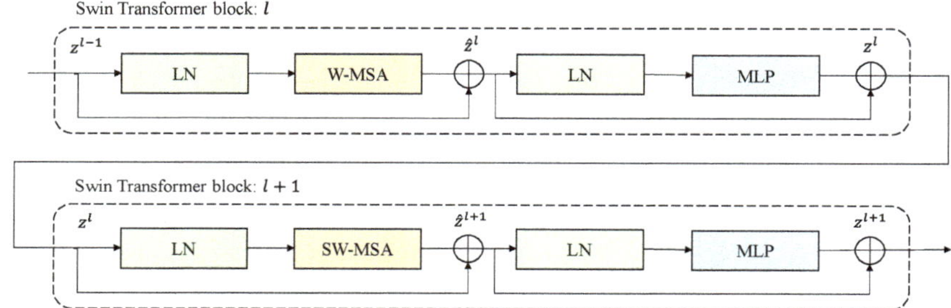

Figure 3. Examples of connections of the Swin transformer blocks.

Connections of Swin transformer blocks can be expressed in Equations (2)–(5) below, where '$\hat{z}^l$' is the output of (S)W-MSA, 'z^l' is the output of MLP, and 'l' denotes the position of the Swin transformer block.

$$\hat{z}^l = W - MSA\left(LN\left(z^{l-1}\right)\right) + z^{l-1} \tag{2}$$

$$z^l = MLP(LN(\hat{z})) + \hat{z}^l \tag{3}$$

$$\hat{z}^{l+1} = SW - MSA\left(LN\left(z^l\right)\right) + z^l \tag{4}$$

$$z^{l+1} = MLP\left(LN\left(\hat{z}^{l+1}\right)\right) + \hat{z}^{l+1} \tag{5}$$

2.3. Past Studies on Models Constructed Based on the ATLAS Dataset

Qi et al. [17] proposed an X-Net with an encoder–decoder structure using 2D CNN to analyze the ATLAS data. The X-block used in X-Net consisted of three (3 × 3) depthwise-separable convolution layers, a 1 × 1 convolution layer, and one 1 × 1 convolution layer that connects input and output. The size of the input and output of X-Net was 224 × 192, and when training the model, the sum of Dice loss and cross-entropy loss was used in the loss function. In the experimental results using the five-fold cross-validation method with the ATLAS dataset, the following performances were obtained: a Dice of 0.4867, IoU of 0.3723, precision of 0.6, and recall of 0.4752.

Zhou et al. [18] proposed a dimension-fusion-UNet (D-UNet) of an encoder–decoder structure that combined 2D and 3D CNNs. Zhou et al. [2] combined four grayscale images into one 3D image, where the inputs of D-UNet were 192 × 192 × 4 in 2D form and 192 × 192 × 4 × 1 in 3D form. The output of D-UNet was 192 × 192 × 1, which was based on the ground truth value corresponding to the third MRI scan image in the four grayscale images used for the generation of a 3D image. D-UNet was trained using the MRI scan images of 183 patients in the ATLAS dataset and validated using the MRI scan images of 46 patients, obtaining a Dice of 0.5349.

Basak et al. [19] modified the CNN decoder in the D-UNet's decoder into a parallel partial decoder (PPD) and the obtained performances were Dice, IoU, precision, and recall of 0.5457, 0.4015, 0.6371, and 0.4969, respectively. Zhang et al. [20] obtained good performance by preprocessing the ATLAS data using the large deformation diffeomorphic metric mapping (LDDMM) method and inputting them into U-Net. Furthermore, they compared the performance of 2D U-Net and 3D U-Net with the same dataset and obtained the following results: when the data without preprocessing was used, the Dice of 2D U-Net was 0.4554, and that of 3D U-Net was 0.5296; when the preprocessed dataset was used, the Dice of 2D U-Net was 0.4954, and that of 3D U-Net was 0.5672. Therefore, the model's performance improved when the ATLAS data were preprocessed using the LDDMM method, and the performance of 3D U-Net was relatively better than that of 2D U-Net. However, it was appropriate to conclude that a 3D model is better than a 2D model in MRI image analysis because, in the experiment of [4], 3D U-Net received a $49 \times 49 \times 49 \times 1$ image as input, whereas 2D U-Net received a $233 \times 197 \times 1$ image, showing significant differences in the layer depth and parameter settings between the two models.

3. ATLAS Dataset

The ATLAS dataset is a standardized open dataset built to train and test algorithms for segmenting lesions of strokes and compare performance [21,22]. The ATLAS dataset was created by collecting 189 MRI scan images with a resolution of 197×233 pixels from 229 patients. Thus, in the ATLAS dataset, there are 189 MRI scans (which are 3D scans) and 43,281 slices are annotated with two classes: normal pixels, and pixels with a disease. In this study, the MRI scan images of 177 patients were used (177×189 slices), where 80% of the total data was randomly assigned as training data and the data of the remaining 52 patients was designated as validation data (52×189 slices).

4. Proposed Method: STHarDNet

4.1. HarDNet

HarDNet is a model of U-Net form with an encoder–decoder structure built with the HarDNet block as a backbone. Figure 4 shows the structure of the HarDNet model. The encoder refers to a process of extracting a feature map while reducing the image size through the down-sampling process and the encoder consists of one convolution block, four HarDNet blocks, and four down-sampling blocks. The convolution block consists of a convolution layer where filter = 24, kernel size = 3, and stride = 2, and a convolution layer where filter = 48, kernel size = 3, and stride = 1. A down-sampling block consists of a convolution layer with kernel size = 1 and an AvgPoll2d layer with kernel size = 2 and stride = 2. The transition section (bottleneck) uses a HarDNet block to complete the parameter transfer of the encoder and the decoder. The decoder up-samples the feature map received from the bottleneck into the same size as the input and, at the same time, it finds the disease region in the feature map and displays it in the output image. The decoder outputs the final image of (W, H) size after going through four HarDNet blocks, five up-sampling blocks, and the last convolution layer with kernel size = 1. An up-sampling block consists of an interpolate function that uses the "bilinear" mode and a convolution layer with kernel size = 1.

If an input image is expressed in terms of (W (width), H (height), and C (channels)), the shape of the feature map calculated through the encoder's convolution block is (W/2, H/2, 48). The feature map of (W/32, H/32, 286) is output through four HarDNet blocks and four down-sampling blocks. The decoder receives the feature map of (W/32, H/32, 320) output from the bottleneck as input and then outputs a feature map of (W, H, class) (the class refers to the number of disease categories in the data). Table 1 shows the HarDNet's detailed structure and output examples, where the output size column shows examples where a grayscale image with a size of 224×224 is used as input.

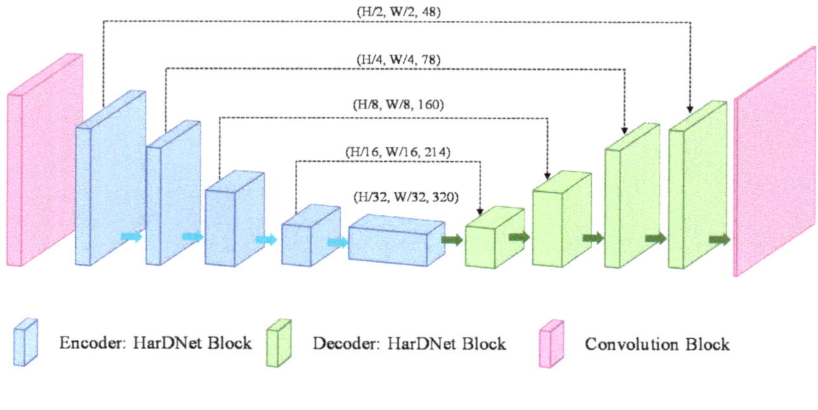

Figure 4. Architecture of HarDNet.

Table 1. HarDNet's structure and output examples.

Stage	Block Name	Details	Output Size
Input			$224 \times 224 \times 1$
Encoder	Conv Block	$2 \times$ Convolution	$112 \times 112 \times 48$
	HarDNet Block	$4 \times$ Convolution	$112 \times 112 \times 48$
	Down-sampling Block	$1 \times$ Convolution $1 \times$ AvgPool2d	$56 \times 56 \times 64$
	HarDNet Block	$4 \times$ Convolution	$56 \times 56 \times 78$
	Down-sampling Block	$1 \times$ Convolution $1 \times$ AvgPool2d	$28 \times 28 \times 96$
	HarDNet Block	$8 \times$ Convolution	$28 \times 28 \times 160$
	Down-sampling Block	$1 \times$ Convolution $1 \times$ AvgPool2d	$14 \times 14 \times 160$
	HarDNet Block	$8 \times$ Convolution	$14 \times 14 \times 214$
	Down-sampling Block	$1 \times$ Convolution $1 \times$ AvgPool2d	$7 \times 7 \times 286$
Bottle Neck	HarDNet Block	$8 \times$ Convolution	$7 \times 7 \times 320$
Decoder	Up-sampling Block	$1 \times$ Upsample $1 \times$ Convolution	-
	HarDNet Block	$8 \times$ Convolution	$14 \times 14 \times 214$
	Up-sampling Block	$1 \times$ Upsample $1 \times$ Convolution	-
	HarDNet Block	$8 \times$ Convolution	$28 \times 28 \times 160$
	Up-sampling Block	$1 \times$ Upsample $1 \times$ Convolution	-
	HarDNet Block	$4 \times$ Convolution	$56 \times 56 \times 78$
	Up-sampling Block	$1 \times$ Upsample $1 \times$ Convolution	-
	HarDNet Block	$4 \times$ Convolution	$112 \times 112 \times 48$
	Up-sampling Block	$1 \times$ Upsample $1 \times$ Convolution	-
Output	Conv Block	$1 \times$ Convolution	$224 \times 224 \times 2$

4.2. Swin Transformer Block

Figure 5 shows the structure of the Swin Transformer model used in this study. The input image passes through the patch partition layer and is segmented into patches with a 4×4 size to generate patch tokens having a shape of (W/4, H/4, $4 \times 4 \times$ channel). The generated patch tokens go through the linear embedding in stage 1. Afterward, they are input into two connected Swin transformer blocks to generate tokens of (W/4, H/4, C) where C refers to an arbitrary dimension. Stages 2 and 3 consist of patch merging and Swin transformer blocks, respectively. In patch merging, adjacent 2×2 patches are merged into one patch, and the tokens are down-sampled to 1/2, whereas C is doubled. In stages 2 and 3, the shape of the tokens is (W/8, H/8, 2C) and (W/16, H/16, 4C), respectively. In the Swin Transformer's last "Linear Projection", the feature dimension is expanded to eight times the input dimension, and after going through a 1×1 convolution, a feature map of the same shape is output. The Swin Transformer model receives a $112 \times 112 \times 48$ feature map that was output from the first HarDNet block of the encoder, which is then connected to the last HarDNet block of the decoder. The window size of the Swin Transformer model used in this study is seven.

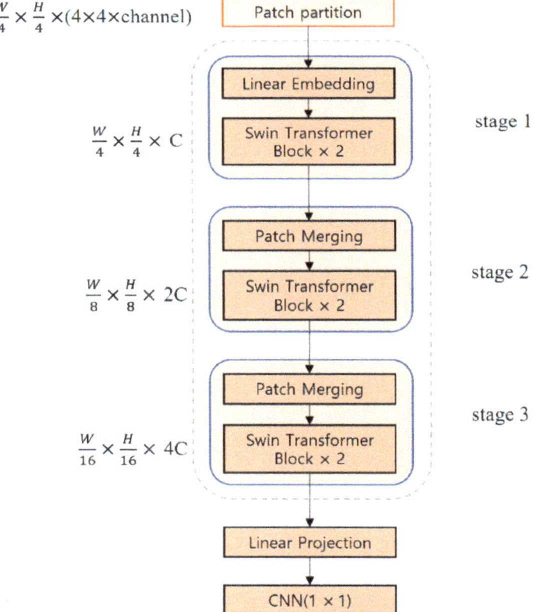

Figure 5. Architecture of the Swin Transformer.

4.3. STHArDNet: Combination of Swin Transformer with HarDNet

To combine the convolution with the vision transformer, this study proposes a STHarD-Net model structure, as shown in Figure 6. STHarDNet consists of (1) the HarDNet with the encoder–decoder structure of the U-Net form and (2) the Swin Transformer used in skip connection that connects the encoder and the decoder.

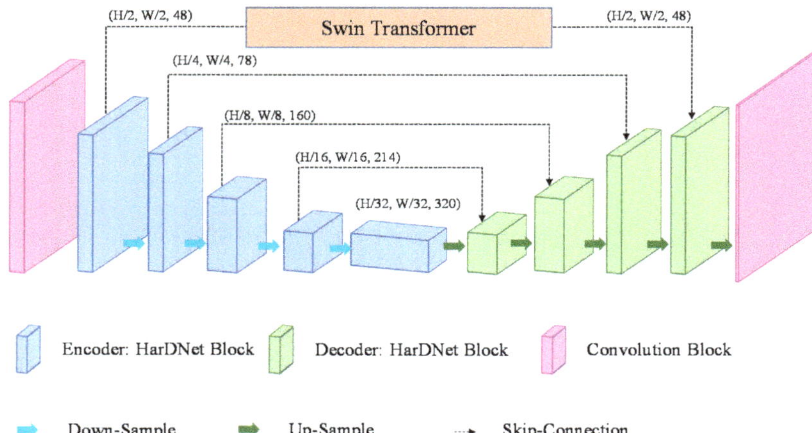

Figure 6. Architecture of the proposed STHarDNet model.

The Swin Transformer uses a hierarchical transformer to extract hierarchical feature maps and uses the shifted windows approach to calculate the relationship between patches in the whole feature map. Therefore, it has a more suitable model structure for segmentation than the transformer. However, the hierarchical transformer down-samples the feature map's size by merging adjacent 2 × 2 patches to generate a hierarchical feature map. Therefore, if the input feature map's size is small, the Swin Transformer cannot extract a deep hierarchical feature map.

Conventional combined models of CNN and transformer used the transformer in the bottleneck of the encoder and the decoder; however, the bottleneck has a small size (7 × 7) because its input is the feature map output at the end of the encoder. This is not suitable for the purpose of using the hierarchical transformer of the Swin Transformer in this study. Therefore, the Swin Transformer was not used at the bottleneck of the encoder and the decoder in this study, but used it in the first skip connection, where the size of the feature map was the largest among skip connections.

5. Experiments

5.1. Performance Evaluation Method

This study used four metrics—Dice, IoU (Intersection over Union, Jaccard index), precision, and recall—to evaluate the model's performance. Dice is a metric used to measure the similarity between the predicted and actual values, and its value ranges from 0 to 1. The calculation equation is the same as that of F1-score, but there is a tendency to emphasize Dice more in the medical image segmentation field. Dice is also expressed as Dice coefficient, Dice similarity coefficient (DSC), and Dice score, depending on papers. IoU is a performance metric used in object detection and semantic segmentation studies and refers to the ratio of the intersection area to the union area for the predicted values and the actual values. IoU has the same meaning as Jaccard index. Precision refers to the percentage of the pixels predicted accurately in the prediction result of the model. Recall refers to how well the model detected the ground truths.

Dice, IoU, precision, and recall can all be calculated using True Positive (TP), True Negative (TN), False Positive (FP), and False Negative (FN) of the classification confusion matrix. Equations (6)–(9) show the calculation methods of the performance metrics, respectively.

$$Dice = \frac{2TP}{2TP + FP + FN} \tag{6}$$

$$IoU = \frac{TP}{TP + FP + FN} \tag{7}$$

$$Precision = \frac{TP}{TP + FP} \tag{8}$$

$$Recall = \frac{TP}{TP + FN} \tag{9}$$

As explained in Section 3, the ATLAS dataset has 3D MRI scan images, and the model's performance was evaluated by measuring the performance at the patient level. In other words, instead of measuring the performance in the prediction result of one image, the disease of a patient was predicted using 189 MRI scan images captured for the patient as the input values of the model. In this study, the data of 52 patients were used as the validation data and obtained a total of 52 prediction results, which were then averaged

5.2. Experimental Setup and Parameter Settings

Table 2 shows the experimental setup used in this study. In the training process, the size of the input images in every model was set to 224 × 224 and the batch size was set to 16. Adam was used as an optimization function. The initial learning rate was set to 0.001, and if the validation loss did not decrease in five epoch cycles, the learning rate was decreased by 0.2 times. If the validation loss did not decrease in ten epoch cycles, early stopping was executed. As regards the loss function, the sum of the Dice loss and cross-entropy was used with the same weight.

Table 2. Experimental setup.

Device	Specifications
OS	Windows 10
CPU	Intel Core i9-9900KF 3.6GHz
GPU	NVIDIA GeForce RTX 2080Ti × 1
RAM	112GB
Storage	1TB SSD
Language	Python 3.7, PyTorch = 1.5

5.3. Performance Comparison Experiments

To validate the model proposed in this study, performance comparison experiments were conducted after selecting nine models that showed good performance in SOTA methods and the ATLAS dataset in past semantic segmentation studies. The SOTA models used in the experiments were typical U-Net-based models (U-Net, SegNet, PSPNet, and HarDNet) in the segmentation field with a CNN-based encoder–decoder structure; segmentation models (Swin Transformer and Swin UNet) using vision transformers; a model (TransHarDNet) that combined the CNN and vision transformer; and X-Net and D-UNet constructed to analyze the ATLAS data.

Table 3 shows the experimental results. The performances shown in Table 3 were obtained from the validation dataset after training the model with a separated dataset. In Table 3, the column "input type" refers to the input image shape of the model, "2D" refers to 2D grayscale images, and "3D" refers to 3D grayscale images. A 3D image was created by combining consecutive MRI scan images and the "output type" of every model was a 2D image. The 2D output corresponding to a 3D image input was based on the target value of the third image in the four consecutive MRI scan images that produced the 3D image and was the same as the 3D input/output used in D-UNet [2].

Table 3. Performance comparison of models in the ATLAS dataset.

Model Name	Input Type	Output Type	Dice	IoU	Precision	Recall
U-Net [4]	2D	2D	0.4517	0.3333	0.4831	0.5118
SegNet [23]	2D	2D	0.3751	0.2675	0.4418	0.3767
PSPNet [24]	2D	2D	0.4465	0.3282	0.5653	0.4200
HarDNet [14]	2D	2D	0.5066	0.3774	0.7358	0.4331
TransHarDNet	2D	2D	0.5051	0.3816	0.5785	0.5176
Swin Transformer [15]	2D	2D	0.1640	0.1053	0.7306	0.1247
Swin UNet [16]	2D	2D	0.4034	0.2883	0.5871	0.3402
X-Net [17]	2D	2D	0.4859	0.3670	0.6277	0.4391
D-UNet [18]	2D+3D	2D	0.4759	0.3570	0.4780	0.5248
STHarDNet	2D	2D	0.5170	0.3866	0.6222	0.4979
STHarDNet (Proposed)	3D	2D	0.5547	0.4184	0.6764	0.5286

As shown in Table 3, the STHarDNet proposed in this study resulted in a Dice of 0.517, IoU of 0.387, precision of 0.622, and recall of 0.498 when the "input type" and "output type" were both 2D. These performances were higher than those of existing SOTA models and stroke diagnostic models. Furthermore, when a 3D image was input and a 2D image was output, the proposed STHarDNet showed a Dice of 0.5547, IoU of 0.4184, precision of 0.6763, and recall of 0.5286, showing higher performances compared to when a 2D image was input. (The shape of the 3D input image of STHarDNet is 224 × 224 × 4).

The experimental results show that the Dice, IoU, precision, and recall performances of the HarDNet built based on a single CNN or a single Swin Transformer-based segmentation model were lower compared to those when the two models are all combined and used. This proves that, if the ViT-based Swin Transformer model and the CNN-based HarDNet model proposed in this study are combined, the performance can be improved compared to when a single model is used.

5.4. Speed Comparison Experiments of Models

The golden time for strokes from the onset to diagnosis and treatment was less than one hour. Therefore, not only the segmentation performance, but also the image process speed, are critically important for stroke diagnosis models. Therefore, comparative experiments of image processing speed were conducted with the STHarDNet and the models used in the performance comparison experiments. In the experiments, the time consumed and frames per second (FPS) were recorded when 100,000 images were processed, respectively, in the same environment. Table 4 shows the results of the experiments on the speed comparison of the models.

Table 4. Speed comparison of models.

Model Name	Input Image Size	100,000 Frames	
		Seconds	FPS
U-Net	224 × 224	475.109	210.477
SegNet	224 × 224	415.869	240.459
PSPNet	224 × 224	637.318	156.907
HarDNet	224 × 224	325.117	307.58
TransHarDNet	256 × 256	352.066	284.036
Swin Transformer	224 × 224	1353.657	73.873
Swin UNet	224 × 224	414.798	241.081
X-Net	224 × 192	982.462	101.785
D-UNet	192 × 192	572.719	174.605
STHarDNet (Proposed)	224 × 224	333.395	299.943

In Table 4, HarDNet showed the fastest speed with 307.58 FPS when processing 100,000 images, whilst STHarDNet was the second fastest with FPS of 299.943 FPS. STHarD-Net was 2.48% slower than HarDNet when performing calculations with 100,000 images, but the Dice and IoU performances were 9.49% and 10.89% better, respectively. Furthermore, when 100,000 images were calculated, the FPS was 42.5% faster than that of the U-Net, 71.78% faster than that of the D-UNet, and 5.6% faster than that of the TransHarDNet that used the transformer in the bottleneck of the HarDNet.

6. Conclusions

This study proposed a STHarDNet structure by combining the Swin Transformer and HarDNet and applied it to the segmentation of stroke MRI scan images. The STHarDNet consists of two models: an encoder–decoder model in a lightweight U-Net shape that consists of HarDNet blocks and a Swin Transformer model consisting of Swin transformer blocks that connect the encoder and the decoder in the first skip connection. STHarDNet has both the character of the CNN model that completes the task while simultaneously looking at the constraining parts, and the character of the transformer that completes the task while looking at the sequence data in all images. By applying the Swin Transformer to the first skip connection in the model, it can receive a feature map larger than the bottleneck as an input, thereby maintaining the advantage of hierarchical feature extraction and the shifted windows approach of the Swin Transformer.

To prove the superiority of the proposed model, the MRI scan images of 177 patients from the ATLAS dataset were used for training and the images of 52 patients were used for validation to conduct the performance comparison experiments with existing SOTA models in the segmentation field. When $224 \times 224 \times 1$ grayscale images were used in the experiments, STHarDNet showed a Dice of 0.517, IoU of 0.387, precision of 0.622, and recall of 0.498, which was better than those of the existing SOTA models used in the experiment. Furthermore, when $224 \times 224 \times 4$ 3D images that gathered four consecutive grayscale images were entered as inputs, the proposed model showed a Dice of 0 0.5547, IoU of 0.4185, precision of 0.6764, and recall of 0.5286, showing higher performances compared to when 2D images were input. Moreover, when 100,000 images were processed in the comparative experiment of image processing speed with the existing SOTA models, the STHarDNet model achieved 299.943 FPS, which was 2.48% slower than HarDNet and the second fastest among the ten models. However, because the segmentation performance of STHarDNet was 9.49% higher in terms of the Dice and 10.89% higher in terms of IoU compared to that of HarDNet, it can be said that the proposed STHarDNet is the best when the performances and speed were both considered.

Through the experiments in this study, the excellent segmentation performance of STHarDNet that combined the Swin Transformer in the first skip connection of HarDNet was demonstrated. Furthermore, it was proven that, even if the transformer is connected to not only the bottleneck, but also the skip connection, the model's performance could be enhanced while maintaining a fast calculation speed.

In this study, the model was developed based on 2D layers. The input 3D image was also analyzed with the 2D layers. In future research, we plan to modify the model with 3D layers to improve its performance.

Author Contributions: Conceptualization, Z.P. and Y.G.; methodology, Z.P.; software, Z.P.; validation, Z.P., Y.G. and S.J.Y.; formal analysis, Y.G. and S.J.Y.; investigation, Y.G.; resources, Y.G.; data curation, Z.P.; writing—original draft preparation, Z.P.; writing—review and editing, Y.G.; visualization, Y.G.; supervision, S.J.Y.; project administration, Y.G.; funding acquisition, Y.G. All authors have read and agreed to the published version of the manuscript.

Funding: This work was supported by the Institute of Information & Communications Technology Planning & Evaluation (IITP) grant funded by the Korean government (MSIT) (No.2021-0-00755/20210007550012002, Dark data analysis technology for data scale and accuracy improvement).

Institutional Review Board Statement: Not applicable.

Informed Consent Statement: Not applicable.

Data Availability Statement: There were no human subjects within this work because of the secondary use of public and de-identified subjects. The patient data for the ATLAS dataset [21] used in this work can be found at the following: http://fcon_1000.projects.nitrc.org/indi/retro/atlas.html (accessed on 15 February 2021).

Conflicts of Interest: The authors declare no conflict of interest. The funders had no role in the design of the study; in the collection, analyses, or interpretation of data; in the writing of the manuscript; or in the decision to publish the results.

References

1. Badža, M.; Barjaktarović, M. Segmentation of Brain Tumors from MRI Images Using Convolutional Autoencoder. *Appl. Sci.* **2021**, *11*, 4317. [CrossRef]
2. Ghosh, S.; Huo, M.; Shawkat, M.S.A.; McCalla, S. Using Convolutional Encoder Networks to Determine the Optimal Magnetic Resonance Image for the Automatic Segmentation of Multiple Sclerosis. *Appl. Sci.* **2021**, *11*, 8335. [CrossRef]
3. Wu, S.; Wu, Y.; Chang, H.; Su, F.T.; Liao, H.; Tseng, W.; Liao, C.; Lai, F.; Hsu, F.; Xiao, F. Deep Learning-Based Segmentation of Various Brain Lesions for Radiosurgery. *Appl. Sci.* **2021**, *11*, 9180. [CrossRef]
4. Ronneberger, O.; Fischer, P.; Brox, T. U-Net: Convolutional Networks for Biomedical Image Segmentation. In Proceedings of the International Conference on Medical Image Computing and Computer-Assisted Intervention, Munich, Germany, 5–9 October 2015; Springer: Cham, Switzerland, 18 November 2015; pp. 234–241.
5. Wang, D.; Wu, Z.; Yu, H. TED-Net: Convolution-Free T2T Vision Transformer-Based Encoder-Decoder Dilation Network for Low-Dose CT Denoising. *arXiv* **2021**, arXiv:2106.04650.
6. Zhou, H.-Y.; Guo, J.; Zhang, Y.; Yu, L.; Wang, L.; Yu, Y. nnFormer: Interleaved Transformer for Volumetric Segmentation. *arXiv* **2021**, arXiv:2109.03201.
7. Carion, N.; Massa, F.; Synnaeve, G.; Usunier, N.; Kirillov, A.; Zagoruyko, S. End-to-End Object Detection with Trans-formers. In *European Conference on Computer Vision*; Springer: Cham, Switzerland, 2020; Volume 12346, pp. 213–229. [CrossRef]
8. Zheng, S.; Lu, J.; Zhao, H.; Zhu, X.; Luo, Z.; Wang, Y.; Fu, Y.; Feng, J.; Xiang, T.; Torr, P.H.; et al. Rethinking Semantic Segmentation from a Sequence-to-Sequence Perspective with Transformers. In Proceedings of the 2021 IEEE/CVF Conference on Computer Vision and Pattern Recognition (CVPR), Nashville, TN, USA, 21–24 June 2021; pp. 6877–6886.
9. Dosovitskiy, A.; Beyer, L.; Kolesnikov, A.; Weissenborn, D.; Zhai, X.; Unterthiner, T.; Dehghani, M.; Minderer, M.; Heigold, G.; Gelly, S.; et al. An Image Is Worth 16x16 Words: Transformers for Image Recognition at Scale. *arXiv* **2020**, arXiv:2010.11929.
10. Xu, G.; Wu, X.; Zhang, X.; He, X. LeViT-UNet: Make Faster Encoders with Transformer for Medical Image Segmentation. *arXiv* **2021**, arXiv:2107.08623.
11. Chang, Y.; Menghan, H.; Guangtao, Z.; Xiao-Ping, Z. TransClaw U-Net: Claw U-Net with Transformers for Medical Image Segmentation. *arXiv* **2021**, arXiv:2107.05188.
12. Chen, J.; Lu, Y.; Yu, Q.; Luo, X.; Adeli, E.; Wang, Y.; Lu, L.; Yuille, A.L.; Zhou, Y. TransUNet: Transformers Make Strong Encoders for Medical Image Segmentation. *arXiv* **2021**, arXiv:2102.04306.
13. Wang, W.; Chen, C.; Ding, M.; Yu, H.; Zha, S.; Li, J. TransBTS: Multimodal Brain Tumor Segmentation Using Transformer. In *Lecture Notes in Computer Science*; Springer: Cham, Switzerland, 2021; pp. 109–119. [CrossRef]
14. Chao, P.; Kao, C.-Y.; Ruan, Y.; Huang, C.-H.; Lin, Y.-L. HarDNet: A Low Memory Traffic Network. In Proceedings of the 2019 IEEE/CVF International Conference on Computer Vision (ICCV), Seoul, Korea, 27 October–2 November 2019; pp. 3551–3560. [CrossRef]
15. Liu, Z.; Lin, Y.; Cao, Y.; Hu, H.; Wei, Y.; Zhang, Z.; Lin, S.; Guo, B. Swin Transformer: Hierarchical Vision Transformer using Shifted Windows. *arXiv* **2021**, arXiv:2103.14030.
16. Cao, H.; Wang, Y.; Chen, J.; Jiang, D.; Zhang, X.; Tian, Q.; Wang, M. Swin-Unet: Unet-like Pure Transformer for Medical Image Segmentation. *arXiv* **2021**, arXiv:2105.05537.
17. Qi, K.; Yang, H.; Li, C.; Liu, Z.; Wang, M.; Liu, Q.; Wang, S. X-Net: Brain Stroke Lesion Segmentation Based on Depthwise Separable Convolution and Long-Range Dependencies. In Proceedings of the International Conference on Medical Image Computing and Computer-Assisted Intervention, Shenzhen, China, 13–17 October 2019; Springer: Cham, Switzerland, 2019; Volume 11766, pp. 247–255. [CrossRef]
18. Zhou, Y.; Huang, W.; Dong, P.; Xia, Y.; Wang, S. D-UNet: A Dimension-Fusion U Shape Network for Chronic Stroke Lesion Segmentation. *IEEE/ACM Trans. Comput. Biol. Bioinform.* **2019**, *18*, 940–950. [CrossRef] [PubMed]
19. Basak, H.; Hussain, R.; Rana, A. DFENet: A Novel Dimension Fusion Edge Guided Network for Brain MRI Segmentation. *SN Comput. Sci.* **2021**, *2*, 1–11. [CrossRef]
20. Zhang, Y.; Wu, J.; Liu, Y.; Chen, Y.; Wu, E.X.; Tang, X. MI-UNet: Multi-Inputs UNet Incorporating Brain Parcellation for Stroke Lesion Segmentation From T1-Weighted Magnetic Resonance Images. *IEEE J. Biomed. Health Inform.* **2021**, *25*, 526–535. [CrossRef] [PubMed]
21. Anatomical Tracings of Lesions after Stroke (ATLAS) R1.1. Available online: http://fcon_1000.projects.nitrc.org/indi/retro/atlas.html (accessed on 15 February 2021).

22. Liew, S.-L.; Anglin, J.M.; Banks, N.W.; Sondag, M.; Ito, K.; Kim, H.; Chan, J.; Ito, J.; Jung, C.; Khoshab, N.; et al. A large, open source dataset of stroke anatomical brain images and manual lesion segmentations. *Sci. Data* **2018**, *5*, 180011. [CrossRef] [PubMed]
23. Badrinarayanan, V.; Kendall, A.; Cipolla, R. SegNnet: A deep convolutional encoder-decoder architecture for image segmentation. *IEEE Trans. Pattern Anal. Mach. Intell.* **2017**, *39*, 2481–2495. [CrossRef] [PubMed]
24. Zhao, H.; Shi, J.; Qi, X.; Wang, X.; Jia, J. Pyramid scene parsing network. In Proceedings of the 2017 IEEE Conference on Computer Vision and Pattern Recognition, Honolulu, HI, USA, 21–26 July 2017; pp. 2881–2890. [CrossRef]

Article

A Whole-Slide Image Managing Library Based on Fastai for Deep Learning in the Context of Histopathology: Two Use-Cases Explained

Christoph Neuner [1], Roland Coras [1], Ingmar Blümcke [1], Alexander Popp [1], Sven M. Schlaffer [2], Andre Wirries [3], Michael Buchfelder [2] and Samir Jabari [1,*]

[1]   Institute of Neuropathology, University Hospital Erlangen, 91054 Erlangen, Germany; christoph.neuner@fau.de (C.N.); roland.coras@uk-erlangen.de (R.C.); Ingmar.Bluemcke@uk-erlangen.de (I.B.); alexander.popp@fau.de (A.P.)

[2]   Department of Neurosurgery, University Hospital Erlangen, 91054 Erlangen, Germany; sven.schlaffer@uk-erlangen.de (S.M.S.); michael.buchfelder@uk-erlangen.de (M.B.)

[3]   Spine Centre, Hessing Foundation, Hessingstrasse 17, 86199 Augsburg, Germany; andre.wirries@hessing-stiftung.de

*   Correspondence: samir.jabari@fau.de

Citation: Neuner, C.; Coras, R.; Blümcke, I.; Popp, A.; Schlaffer, S.M.; Wirries, A.; Buchfelder, M.; Jabari, S. A Whole-Slide Image Managing Library Based on Fastai for Deep Learning in the Context of Histopathology: Two Use-Cases Explained. *Appl. Sci.* **2022**, *12*, 13. https://doi.org/10.3390/app12010013

Academic Editors: Kyungtae Kang, Hyo-Joong Suh and Junggab Son

Received: 22 October 2021
Accepted: 17 December 2021
Published: 21 December 2021

Publisher's Note: MDPI stays neutral with regard to jurisdictional claims in published maps and institutional affiliations.

Abstract: Background: Processing whole-slide images (WSI) to train neural networks can be intricate and labor intensive. We developed an open-source library dealing with recurrent tasks in the processing of WSI and helping with the training and evaluation of neuronal networks for classification tasks. Methods: Two histopathology use-cases were selected and only hematoxylin and eosin (H&E) stained slides were used. The first use case was a two-class classification problem. We trained a convolutional neuronal network (CNN) to distinguish between dysembryoplastic neuroepithelial tumor (DNET) and ganglioglioma (GG), two neuropathological low-grade epilepsy-associated tumor entities. Within the second use case, we included four clinicopathological disease conditions in a multilabel approach. Here we trained a CNN to predict the hormone expression profile of pituitary adenomas. In the same approach, we also predicted clinically silent corticotroph adenoma. Results: Our DNET-GG classifier achieved an AUC of 1.00 for the ROC curve. For the second use case, the best performing CNN achieved an area under the curve (AUC) of 0.97 for the receiver operating characteristic (ROC) for corticotroph adenoma, 0.86 for silent corticotroph adenoma, and 0.98 for gonadotroph adenoma. All scores were calculated with the help of our library on predictions on a case basis. Conclusions: Our comprehensive and fastai-compatible library is helpful to standardize the workflow and minimize the burden of training a CNN. Indeed, our trained CNNs extracted neuropathologically relevant information from the WSI. This approach will supplement the clinicopathological diagnosis of brain tumors, which is currently based on cost-intensive microscopic examination and variable panels of immunohistochemical stainings.

Keywords: brain; pituitary adenoma; dysembryoplastic neuroepithelial tumor; DNET; ganglioglioma; deep learning; digital pathology; convolutional neural network; computer vision; machine learning; convolutional neural network; CNN

1. Introduction

With the increasing availability of digital microscopy scanners and whole slide imaging, digital pathology (DP) will continue to successfully grow into our daily routine diagnostic practice. Whole-slide images, as they are digitized slides, provide the intriguing opportunity for the application of image analysis techniques for advanced tasks, such as disease classification. Deep learning (DL) is the most commonly applied technology in the realm of feature learning. The process involves the iterative improvement of learned representations of regions of interest to achieve maximum class separability. Medical (and nonmedical) image classification tasks have been remarkably successful utilizing DL. The area of computational image analysis of DP images has been already addressed by some previous works. Successful examples range from utilization of different types of cancer detection, classification, or grading [1,2]. Recent work has shown that the differentiation of histologically similar lesions in Focal Cortical Dysplasia in human focal epilepsies is possible [3]. What is more remarkable is that these pathologies differed only in genotype and not in phenotype. Classification of liver cirrhosis, heart failure detection, and classification of Alzheimer's plaques [4] have also been successfully tackled [5]. Lymph node screening to search for metastatic breast cancer has been successfully performed with the help of deep convolutional neuronal networks. Classification of skin lesions has also been successfully performed with the help of DL and elegantly distributed to smartphones for easy daily use of non-expert users [6]. Disease grading, prognosis prediction, and imaging biomarkers for genetic subtype identification are more challenging tasks but have also been successfully established [7–9].

All of these works have shown that deep learning in the context of pathology is becoming more and more common.

However, a prerequisite to successfully apply deep learning requires domain-associated knowledge in the field of DL and DP. Whereas many pathologists are not familiar with the problem-specific tasks and technical issues for applying DL techniques, DL developers most often have little experience with histology and histopathology-associated workflows. In addition, currently available open-source tools and tutorials do not provide guidance for the needs of both groups, and available programming libraries and tools (either open- or closed-source) are not targeted for an application by a pathologist or clinician with little experience in DL programming routine. This is a major obstacle for researchers to use or extend the available technology and investigate their clinical use-case and hypotheses. We developed, therefore, an open-source library specifically tuned and adjusted to the special needs of digital pathology-associated analysis tasks in the context of DL. We showcase the potential of our library by outlining two specific projects, each driven by a unique clinical hypothesis.

1.1. Use Case 1: Classifying Low-Grade Epilepsy-Associated Brain Tumors

Dysembryoplastic neuroepithelial tumor (DNET) and ganglioglioma (GG) are slowly growing tumors composed of both glial and neuronal cell elements and, histopathologically, are often difficult to classify [10] (see Figure 1).

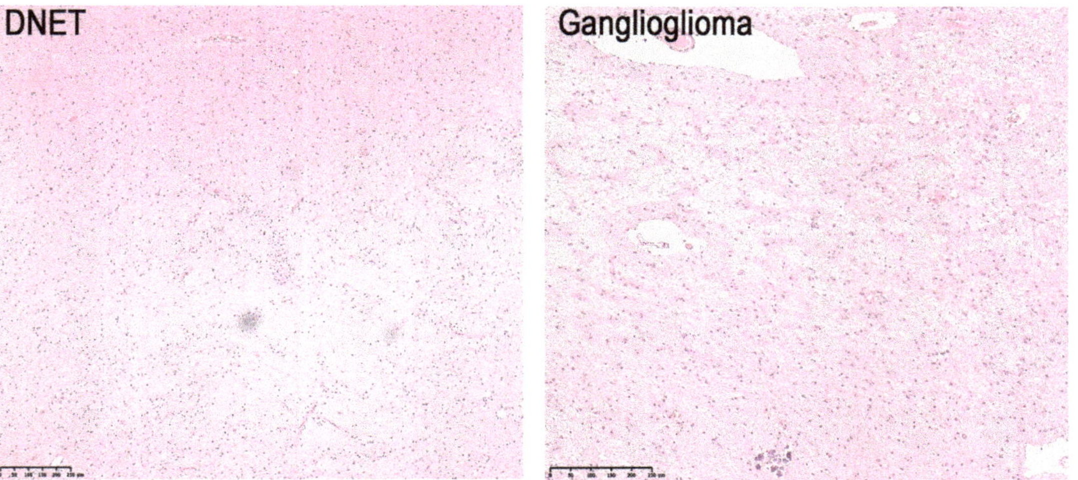

Figure 1. Histopathologic findings in DNET (**left** part) and Ganglioglioma (**right** part). The histomorphological pattern can be hard to tell apart in some cases.

They account for 1–2% of all brain tumors and do not metastasize or spread beyond the primary site of origin. These tumors occur mainly in children and young adults with long-standing drug-resistant epilepsy. The average age at seizure onset was 12 years in 984 GG and 14 years in 565 DNET when reviewing a large European cohort of 9523 patients who underwent epilepsy surgery. Seizures are commonly focal with or without secondary generalization, and neurosurgical resection has proven as the most successful treatment option. Malignant transformation has been reported for the group of GG [11,12], whereas DNET rarely show this behavior [13]. Therefore, a precise histopathological diagnosis and differentiation of these two tumor entities is important for clinical patient management [14]. The problem is that even in specialized medical centers the inter-rater agreement on the diagnosis accounts for only 40% of these tumors [10]. The DL task was to develop, therefore, a binary classifier distinguishing between the two entities.

1.2. Use Case 2: Prediction of Pituitary Adenoma Subtypes and Their Neuroendocrine Features

Better neuroimaging techniques and diagnostic modalities recognize more pituitary adenomas than previously thought [15]. We consider three clinical subclasses: Pituitary adenomas with A. prominent neuroendocrine symptoms, B. slowly developing, insidious, nonspecific complaints delaying accurate diagnosis, or C. incidentally detected adenomas being symptom-free. It remains, therefore, challenging to accurately determine the prevalence and incidence of pituitary adenomas in the general population. They account for 15% of all intracranial neoplasms, being the third most frequent tumor type after meningiomas and gliomas. In multiple postmortem studies, the mean prevalence of pituitary adenomas was 14.4% [15] . The overall estimated prevalence of pituitary adenomas in the general population was calculated as 16.7%. Radiography studies showed a higher prevalence of 22.5% [15,16]. The tumor has its maximum age frequency in patients between 40 and 60 years of age. The frequency of different subtypes varies depending on the age and gender of the patients [16] .

The WHO classification of pituitary adenoma from 2017 is based mainly on the hormone and transcription factor expression of the adenoma cells [17] . In common routine workup for adenomas of the pituitary gland, the morphological evaluation is based, therefore, on H&E and a panel of immunohistochemical staining for all pituitary hormones (adrenocorticotropic hormone (ACTH), luteinizing hormone (LH), follicle-stimulating hormone (FSH), prolactin (PRL), thyroid-stimulating hormone (TSH), and somatotropic hormone (STH)) and transcription factors. In our study, we focused on corticotroph and gonadotroph adenomas (see Figure 2) since they represent the most common subtypes. We labeled our tumor samples of corticotroph and gonadotroph adenomas accordingly, e.g., corticotroph adenoma, gonadotroph adenoma with the expression of LH, and gonadotroph adenoma with the expression of FSH. As adenomas are often nonexclusively positive for only one hormone, many cases received more than one label. Therefore, we chose to tackle the problem as a multilabel approach, which means that the different classes are rated and scored individually, and possible correlations must be learned by the CNN. To make sure that the labels are correct for each tile, we manually reviewed the extracted regions from the H&E slides with the corresponding regions in the immunohistochemically stained images. In addition, we included those corticotroph adenomas as a separate class, in which the patient does not show clinical symptoms of Morbus Cushing (silent corticotropic adenoma). The DL tasks were to classify entities of adenomas of the pituitary gland from H&E-stained slides as well as to predict the clinical parameter of asymptomatic or clinically silent corticotroph adenomas.

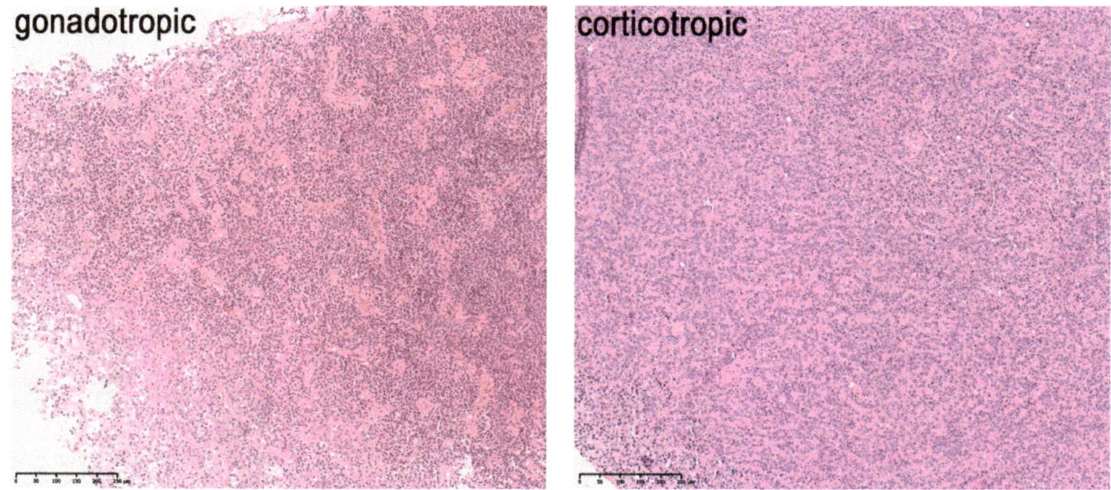

Figure 2. Histopathologic findings in gonadotropic (**left** part) and corticotropic (**right** part) pituitary gland adenoma. A typical pattern in gonadotropic adenoma is the pseudo sinusoidal growth pattern.

What is new:

The depicted library enables users to perform DL with state-of-the-art techniques without the burden of managing WSI-associated overhead, such as pyramid level control or region-specific mapping, as it is kept away from the user. Additionally, the library is fully compatible with one of the most popular deep learning frameworks "fastai" which is based on "PyTorch".

Related work:

In the context of neuropathology-related tasks, few works have been published. Some work has been completed on classifying and detecting Alzheimer's associated lesions, such as extracellular amyloid and intracellular tau deposits [4,18,19]. The latter approach has also been used to classify other tauopathies such as Pick's disease for example [20]. Additionally, with the help of deep learning new disease-correlating features were identified in the white matter of different tauopathies [21]. Classifying glioma and differentiating glioma subtypes from H&E-stained slides and molecular markers was another successful task accomplished [22]. In our own recent project, we could discriminate between phenotypically very similar but genotypically different lesions of focal cortical dysplasia type IIb and tuberous sclerosis complex [3].

2. Materials and Methods

2.1. The Library

Compared to common image datasets consisting of small files in, e.g., PNG or TIFF format, WSI provide more challenges in the context of training a neural network with them. First, there is the size. A WSI's typical size in the realm of Neuropathology is 0.5–3 Gbyte. Therefore, it is impossible to feed an entire WSI let alone a batch of WSI into a CNN, since graphic processing units or graphic cards (GPUs) do not have enough memory. So WSI need to be divided into smaller images usually referred to as tiles. WSI are also stored in special file types and most WSI scanner manufacturers provide their own. Usually, WSI are not independent of each other. A WSI belongs to a case, and a case belongs to a patient. This is important for the dataset split and evaluation of the model after the training. It is common practice to not mix data from one patient in the training, validation, and test set. For evaluation, it is interesting how the model performs on tile level, but usually, the performance on WSI, case or patient-level has a higher value in practice. So, these connections need to be tracked throughout the whole process from preprocessing until postprocessing/evaluation. Our library [23] is meant to help with this common overhead in preprocessing and the evaluation for training a classification model with WSI.

2.2. Tile Calculation

The first step is to split a WSI into multiple small tiles. A complete sample pipeline can be found in the GitHub repository of the library (https://github.com/FAU-DLM/wsi_processing_pipeline/tree/master/tile_extraction/example.ipynb, accessed on 15 December 2021) and the repositories of the two use cases (https://github.com/ChristophNeuner/DNET_vs_Ganglioglioma/blob/main/dnet_vs_gg.ipynb, accessed on 15 December 2021) (https://github.com/ChristophNeuner/glioblastoma_methylation/blob/master/methylation_status_binary_classification.ipynb, accessed on 15 December 2021).

Usually, not all parts of a WSI are of interest for further processing. So, in general, there are two main ways of making sure only the relevant parts are used: marking the interesting regions manually or using some sort of filtering algorithms that, e.g., distinguish tissue from the background, filter out pencil markings, or blurred tissue. Both ways are supported by the library and will be further explained in the following lines.

2.3. Filters Applied on Complete WSI

Our library originated as a fork of Deron Eriksson's GitHub repository "python-wsi-preprocessing" (https://github.com/deroneriksson/python-wsi-preprocessing, accessed on 15 December 2021), which was originally written and used for his and his team's participation in the Tumor Proliferation Assessment Challenge 2016 (TUPAC16) [24].

Most parts of this library have had a substantial rewrite, and many additions were made since. However, the filters were mostly kept untouched. Documentation about them can be found in Deron Erikson's GitHub repository (https://github.com/deroneriksson/python-wsi-preprocessing/blob/master/docs/wsi-preprocessing-in-python/index.md#apply-filters-for-tissue-segmentation, accessed on 15 December 2021) [25] .

2.4. Calculation of Tile Locations

Our preferred way of defining the polygonal regions of interest (ROIs) in a WSI is to use the program QuPath [26] (Supplement S7). The next step is to extract the coordinates of the polygons' vertices. We wrote a small QuPath script that can be used in the "Automate" Tab in QuPath and exports the polygons' vertices' coordinates into a JSON file (https://github.com/FAU-DLM/wsi_processing_pipeline/blob/master/QuPath_scripts/polygon_points_to_json.groovy, accessed on 15 December 2021).

The next step is to convert this information into RegionOfInterestPolygon objects (https://github.com/FAU-DLM/wsi_processing_pipeline/blob/master/shared/roi.py#L66, accessed on 15 December 2021). There is a convenience function if the ROIs were annotated and extracted with our script from QuPath. (https://github.com/FAU-DLM/wsi_processing_pipeline/blob/master/shared/roi.py#L195, accessed on 15 December 2021)

It is important to notice that this part is completely optional. The ROI definition may be skipped.

Subsequently, all relevant tile locations are calculated. For this process, the function "WsisToTilesParallel" (https://github.com/FAU-DLM/wsi_processing_pipeline/blob/8c5e4a360fa369221ce86dd35837e91f31817d30/tile_extraction/tiles.py#L1275, accessed on 15 December 2021) is used. It calls the function "WsiToTiles" for every WSI and runs in parallel. It takes a few interesting parameters. We will elaborate on a few here; the rest is covered in the function's docstring.

"wsi_paths":

First of all, a list with the paths to the WSI files has to be passed. Notice that not only WSI files but also PNG files are supported. If one has already extracted the interesting parts of the WSI as PNGs, one can use them without specifying ROI coordinates, as described before.

"grids_per_roi", "optimize_grid_angles", "angle_stepsize", "minimal_tile_roi_intersection_ratio":

The library lays a grid of all possible tiles over each ROI (Supplement S8). If no ROI is specified, the library internally creates one ROI, which simply spans the complete WSI.

The logic for this part of the pipeline resides in the tiles.py module, to be more specific, in the Vertex, Rectangle, Grid, and GridManager classes. A Vertex object represents one vertex of the polygonal ROI and provides simple arithmetic operations such as add, subtract, and multiply with scalars and matrices. It also provides the functionality to rotate itself around a specified point. This is performed by multiplying a rotation matrix with the vertex coordinates represented as a 2×1 vector.

$$\text{Rotation Matrix } \begin{matrix} x\prime \\ y\prime \end{matrix} = \begin{bmatrix} cos(\alpha) & -sin(\alpha) \\ sin(\alpha) & cos(\alpha) \end{bmatrix} \begin{bmatrix} x \\ y \end{bmatrix}$$

The Vertex class also provides a convenience function to change the WSI level of the coordinates. Because of its size, a WSI is stored in a pyramid-like format (Supplement S10) in multiple images per level. So particular regions of the image are loaded on-demand with higher resolution while zooming in. Therefore during the process of tile calculations, it is important to specify the zoom level for a given coordinate. So, it is often necessary to convert various coordinate values to another zoom level. All the filtering steps for example in our pipeline are performed on a scaled-down version by the factor 32 of the WSI to enhance the speed and obtain the results in a reasonable time.

A Rectangle object represents the bounds of a tile. It also wraps necessary functionality, such as rotation. The Grid class implements all the functionality to represent a grid of Rectangles and, therefore, possible tile locations that are laid over a ROI. Finally, there is the GridManager class. It creates as many Grid objects for each ROI as is specified in "grids_per_roi" and contains some convenience functions for, e.g., visualization. It also merges overlapping ROIs. The full spectrum of the functionality of these classes can be seen on GitHub: https://github.com/FAU-DLM/wsi_processing_pipeline/blob/master/tile_extraction/tiles.py#L78, accessed on 15 December 2021.

If "grids_per_roi" is greater than one, multiple slightly shifted grids are laid over each ROI. This increases the number of tiles and therefore the amount of training data. This means that the same tissue is present in multiple tiles but, nonetheless, all tiles are unique. If "optimize_grid_angles" is true, the grid is rotated in an iterative approach by "angle_stepsize" in each iteration, and the angle, which results in the most tiles per ROI, will be used for further calculations. This is completed for each ROI individually. So the smaller the "angle_stepsize" is, the closer the angle gets to the optimum, but the longer the process takes. The last important parameter in this context is "minimal_tile_roi_intersection_ratio". If it is 1.0, only tiles that lay 100% in the ROI will be considered for further processing. The closer it gets to 0.0, the more tiles can be outside of the ROI, but never completely, since 0.0 is outside of the possible range of this value.

2.5. Tile Filtering

Among these tiles, there might still be some, which are not worth keeping. If ROIs are specified, this amount should be fairly small, but if no ROIs are specified, there should be plenty to be filtered out. The user of the library can specify a tile scoring function that only takes the tile in form of a PIL image as a parameter and returns a score for it. The user also has to provide a threshold for that score. All tiles with a score above this threshold pass filtering and will be considered for training.

The library provides a default tile scoring functionality that works for H&E-stained slides.

$$score = 1 - \cfrac{10}{10 + \cfrac{tissuePercentage * colorFactor * saturationAndValueFactor}{1000}}$$

The scoring formula generates good results for the images in the dataset and was developed through experimentation with the training dataset.

The first criterion is the amount of tissue in a tile. To separate tissue from the background we applied four filters to a tile image (Supplement S9). First, the image was converted to greyscale; then, its complement was created. After that Otsu's threshold was applied. Thresholding using Otsu's method is a popular thresholding technique. This technique was used in the image processing described in A Unified Framework for Tumor Proliferation Score Prediction in Breast Histopathology [27].

The colorFactor value is used to weigh hematoxylin staining heavier than eosin staining. Utilizing the Hue-Saturation-Value (HSV) color model, broad saturation and value distributions are given more weight by the *saturationAndValueFactor*. The *score* is scaled to a value from 0.0 to 1.0.

Tissue with hematoxylin staining is most likely preferable to eosin staining. Hematoxylin stains acidic structures such as DNA and RNA with a purple tone, while eosin stains basic structures such as cytoplasm proteins with a pink tone.

Differentiating purplish shades from pinkish shades can be difficult using the RGB color space [28]. Therefore, to compute the colorFactor value, we first convert the tile's RGB color space to an HSV color space [29]. In this color model, the hue is represented as a degree value on a circle. Purple has a hue of 270 degrees and pink has a hue of 330 degrees. We remove all hues less than 260 and greater than 340. Next, we compute the deviation from purple (270) and the deviation from pink (330). We compute an average factor which is the squared difference of 340 and the hue average. Saturation and value standard deviations should be relatively broad if the tile contains significant tissue. The colorFactor is computed as the pink deviation times the average factor divided by the purple deviation. It favors purple (hematoxylin stained) tissue over pink (eosin stained) tissue. The information about one tile is then stored in a Tile object.

The result of the filtering process is a TileSummary object for each WSI. A TileSummary object contains the information about the WSI including dimensions, scaled dimensions, which were used for faster tile calculations, ROIs, the GridManager object, and all tiles. It also implements some visualization methods to display the WSI with ROI and tile boundaries.

In the next step, the PatientManager class in the wsi_processing_pipeline.shared.patient_manager.py is important. Its main purpose is to manage the hierarchical structure of a pathological dataset. A tile belongs to an ROI. An ROI belongs to a WSI. A WSI belongs to a case, and a case belongs to a patient. It is good practice to split datasets on the patient level. To measure the performance of a model after training, not only can model performance on a tile level be evaluated, but also performance on the WSI or case level is easily assessable. Therefore these relationships are conserved by the PatientManager. It is also responsible for setting the labels of each tile. The PatientManager class additionally implements some convenience functions for dataset splitting into a training, validation, and test set and for a k-fold cross validation split. It can print out a class distribution and is capable of undersampling the dataset.

In the next step, the fastai [30] library takes over for training the neural network. During tile filtering, the user of our library can specify in the WsiToTiles function if each tile should be extracted and stored to disc as a PNG file. We wrote a custom fastai ImageBlock called TileImageBlock that works with fastai's data block API. This allows renouncing saving each tile to disc because the TileImageBlock can extract a tile image on the fly during the training process given the spatial information about a tile that is stored in each Tile object. This has the advantage of consuming less storage space and since it is usually necessary to play around with the parameters that are used for filtering until only the desired tiles are left, not saving the tiles is a huge speedup for this part of the process.

Our preferred library for training a neural network is fastai [30], which is built on top of Facebook's increasingly popular PyTorch [31] library.

After training has finished, evaluating the performance of the model on the validation or an unseen test set is crucial. For this use-case, we implemented the Predictor class, which resides in wsi_processing_pipeline.postprocessing.predictor.py. It takes a fastai [30] Learner and one of our library's PatientManager class objects. In a first step, it calculates predictions for each tile image in the desired dataset. In a second step, it calculates the predictions for each WSI or case by calculating the mean raw prediction for all classes for each tile and applying a threshold that can be specified for each class by the user of the library.

The last step is to evaluate the performance of the model. We, therefore, implemented the Evaluator class in wsi_processing_pipeline.postprocessing.evaluator.py.

Its constructor takes an instance of the abovementioned Predictor class as the only argument. It implements a few commonly used methods to measure model performance. It can calculate the per-class accuracy and plot receiver operating characteristic (ROC) curves, precision-recall curves, confusion matrices (Figure 3), and probability histograms (Figure 4). It can also print out sklearn's classification report and print a list of tiles with the highest losses or a list of cases, WSI, or tiles sorted by a user-specified metric calculated with the predictions. It is also capable of creating Gradient-weighted Class Activation Mappings (Grad-CAMs) [32].

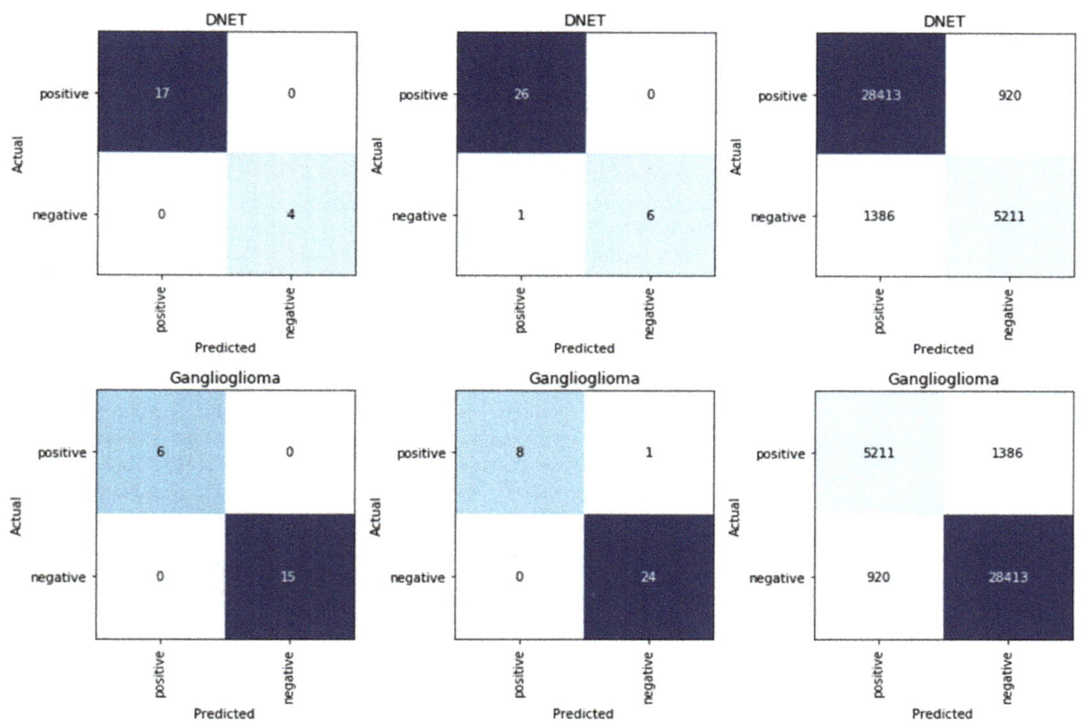

Figure 3. Results Confusion Matrices from left to right: case level, slide level, tile level.

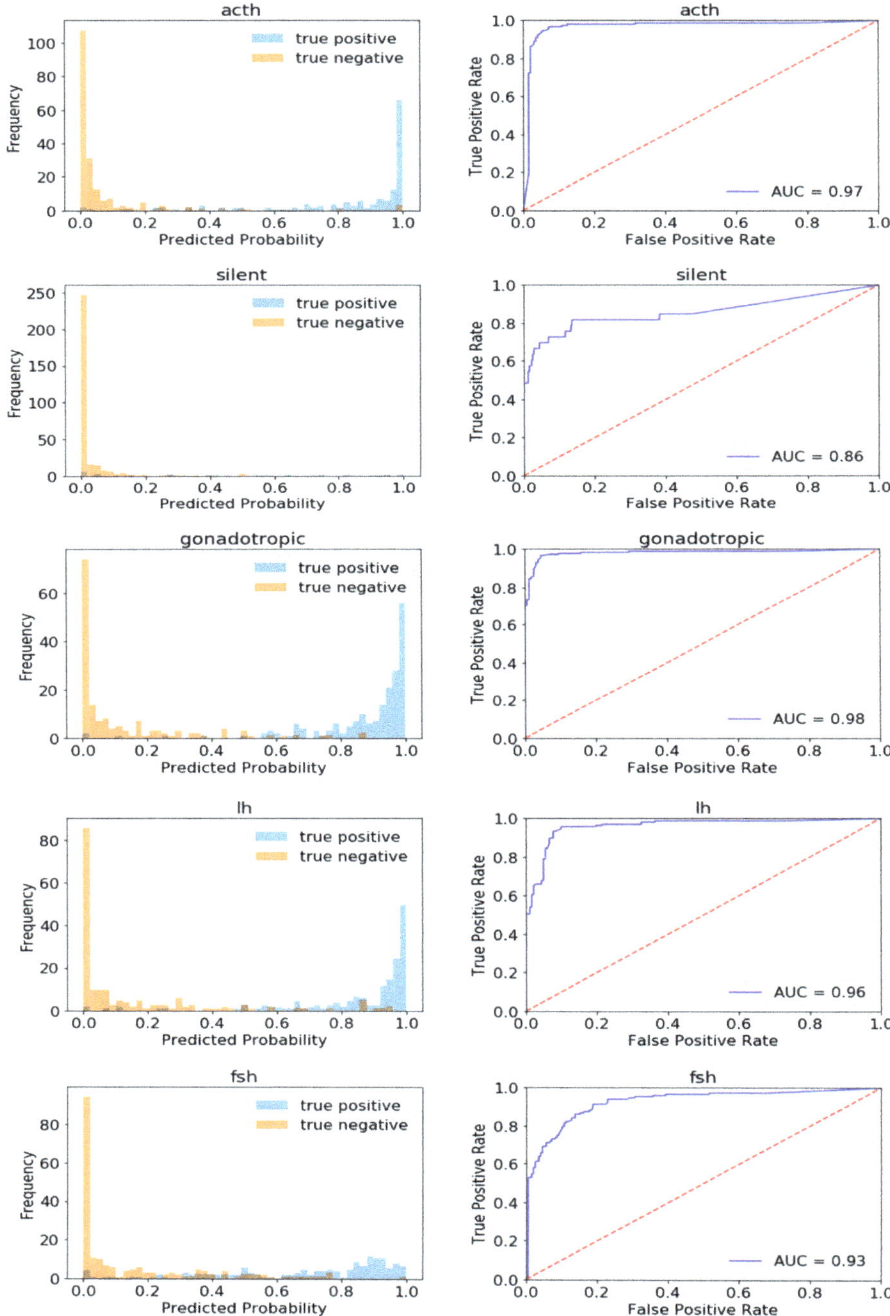

Figure 4. Results | Histograms and ROC-Curves were calculated on a case basis. The predictions were made for all 5 validation sets with the corresponding model that was not trained on that validation set. So, the graphs represent the complete dataset.

2.6. Dataset Preparation for Both Use-Cases

Histopathology slides from all patients of interest for the study design were retrieved from the archives of the Dept. of Neuropathology (see below) and subsequently digitized using a Hamamatsu S60 scanner with a 40× magnification. We included only H&E stainings, thus, eliminating the need for more complex and expensive immunostainings. The WSI of our dataset were reviewed by two expert neuropathologists of our institute.

Use case 1: For the DNET and ganglioglioma, classifier slides from 219 patients were used. In total, 52 of them were DNETs and 167 were ganglioglioma. QuPAth was used by two of our expert neuropathologists in epilepsy pathology to define polygonal ROIs containing tumor tissue in the WSI, and we exported their coordinates to JSON files. These JSON files were then used by the library to extract tiles from the relevant regions of the WSI. In total 171,514 tiles from GG and 34,520 tiles from DNETs with a size of 1024 × 1024 pixels were defined for further processing and training.

Use case 2: To train and evaluate the pituitary adenoma classifier, H&E and immunohistochemically stained (ACTH, LH, FSH) tissue slides of 410 patients were collected. In total, 181 of these were diagnosed with corticotroph and 229 with gonadotroph adenoma of the pituitary gland (Supplement S1 and S2). Overall, the dataset consisted of 431 H&E (202 corticotroph and 229 gonadotroph) slides with the corresponding ACTH LH/FSH whole-slide images for comparing and identifying the correct ROI (Figure 5). The ROIs on an individual H&E slide were defined as regions, where the immunostainings showed tumor expressions of the specific hormone. Care was taken that no normal pituitary gland tissue was included (Figure 5). This time-consuming ROI selection process was necessary to ensure the correct labeling of each tile and, therefore, the validity of the resulting models. Otherwise, biases through wrong labeled areas could have worsened the performance. For example, areas with only connective tissue were excluded. Moreover, the hormone expression of the adenoma is not homogeneously spread over the sample. This was particularly important to consider for gonadotropic adenomas. When an adenoma expresses LH and FSH that does not mean that all subregions express both hormones. So, there can be tiles that are only labeled with LH or FSH, although the whole tumor expresses both. ROIs were defined at 40× magnification level and cropped into smaller tiles of 1024 × 1024 pixels to further preprocess and feed into our model (Figure 5). The tile extraction resulted in 206,517 gonadotropic and 63,893 corticotropic tiles.

2.7. Convolutional Neural Network Architecture

Use case 1: For the DNET-GG classifier, a ResNet50 was implemented, using the open-source Python library fastai [30], which is based on PyTorch [31]. It was pretrained on ImageNet [33,34], and the classification head was replaced to predict two (DNET or GG) instead of the 1000 classes included in the ImageNet dataset (Supplement S3). In our experience, ResNet50 is often a good starting point, since it is relatively fast to train compared to more complex models with more parameters but nonetheless delivers promising results. Since it performed well on the defined dataset, it was not necessary in our view to try out another model.

Use case 2: For the pituitary gland classifier a ResNeXt-101-32x8d CNN architecture also pretrained on ImageNet [33,34] was implemented. ResNeXt-101-32x8d [35,36] was chosen, as it yielded the best results with the least overfitting out of a couple of state-of-the-art network architectures including ResNet50, se_ResNeXt101_32x4d, xception, and inceptionv4 (Supplement S5). The basic network architecture was not changed. Only a customized classification head (Figure 6, Supplement S3) was used to predict four instead of the 1000 ImageNet classes. It consisted of several pooling, batch normalization, dropout, and fully connected layers with four final output channels with a sigmoid-activation function with a threshold of 0.5 to produce individual output probabilities representing the four classes of corticotropic adenoma, silent corticotropic adenoma, gonadotropic adenoma with the expression of LH, and gonadotropic adenoma with the expression of FSH (Figure 6).

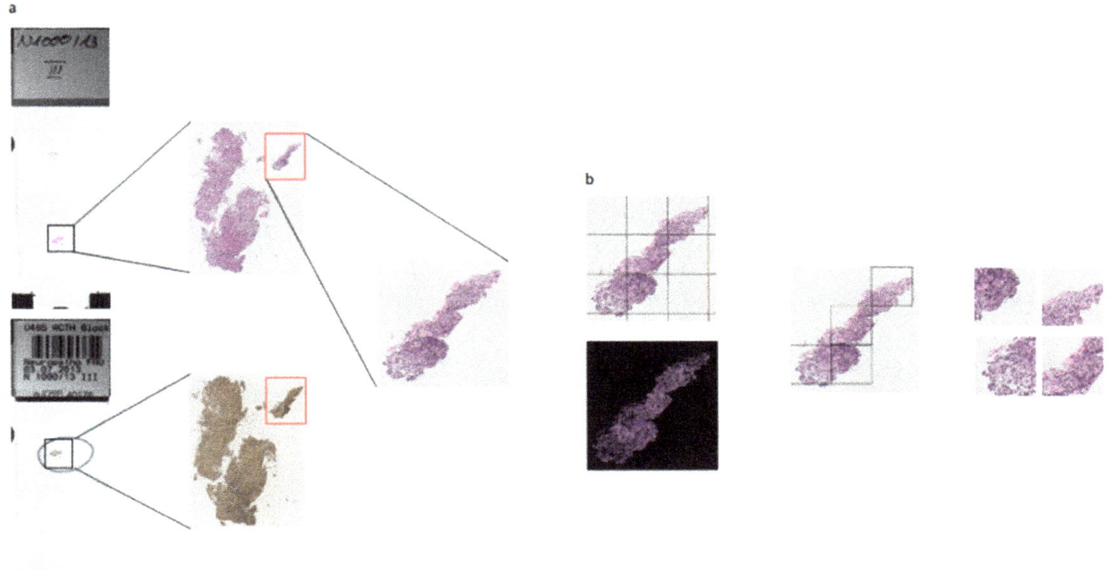

Figure 5. Tile Extraction. (**a**): We compared H&E- and immuno-stained slides and extracted only those corresponding parts of the H&E-stained WSI with QuPath, where the immuno-stained WSI showed the expression of the hormone. (**b**): We subdivided the image into 1024×1024 pixel tiles and used complement filter and otsu thresholds to identify tissue and background. Then we only extracted and saved those tiles that passed a scoring function that takes tissue percentage and color characteristics into account.

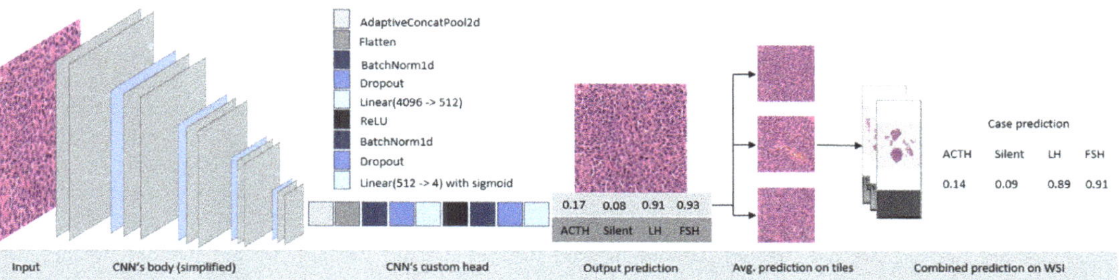

Figure 6. Prediction Pipeline. A tile is forwarded through the model, and the model outputs four independent probabilities for each class. If the probability is over a certain threshold (0.5), the tile obtains the label. All tiles of one case are evaluated, and if more than 50% of the tiles are labeled with one class, the case is also labeled with that class (majority voting).

2.8. Preprocessing and Data Augmentation

Image preprocessing is an important step in every computer vision task to augment the number of samples, to prevent overfitting, and to support the model against invariant aspects that are not correlated with the label [37,38]. First, the tiles were resized to 512×512 pixel images to increase the possible batch size. Following this approach, we made sure to have a wider field of view per tile instead of the maximum possible resolution. In our approach, we used a pipeline of several augmentation techniques performed batch-wise on the GPU consisting of a random crop with reflection padding, randomly flipping (horizontal or vertical), and rotating by a multiple of 90 degrees, a random symmetric warp with a magnitude between -0.2 and 0.2, a random rotation between -10 and $+10$ degrees, a random zoom with a zoom factor between 1.0 and 1.1, and a random change in brightness with a factor between 0.4 and 0.6, where a factor of 0 will transform the image to black, a factor of 1 will transform the image to white, and a factor of 0.5 doesn't adjust the brightness. Furthermore, an augmentation on the contrast of the image was applied with a factor between 0.8 and 1.25, where a factor of 0 will transform the image to grey, a factor over 1 will transform the picture to super-contrast, and a factor = 1 does not adjust the contrast. These augmented images were then normalized. The augmentations were applied on the fly with a randomness factor for reproducibility for every batch so that there was no need to save augmented images and one image could be augmented in multiple ways. This whole approach ensures that out of one image multiple new images of the same class can be obtained by multiplying the number of images available for training the neural network. We tried to apply as little data augmentation as possible to avoid changing special characteristics of the tissue.

2.9. Training and Evaluation

The training was performed with 16-bit precision floating-point numbers [39] using the Adam-Optimizer [40], and the initial learning rate was determined by using fastai's learning rate finder (Supplement S4). The learning rate was adjusted during the training according to the one-cycle policy [41]. The batch size was twelve for the pituitary adenoma classifier and 35 for the DNET-GG classifier. At first, only the randomly initialized custom head (Figure 6, Supplement S3) was trained for five epochs with a maximum learning rate of 10^{-3} (Supplement S4) in both projects to not interfere with the pretrained weights of the CNN's body. Thereafter the body's layers were unfrozen, and the complete network was trained for ten epochs with differential learning rates between 10^{-9} and 10^{-6} for the pituitary gland adenoma classifier and between 10^{-8} and 10^{-6} for the DNET-GG classifier (Supplement S4) where earlier layers were trained with a lower learning rate than the later ones. The idea behind this is to maintain the basic image-classification patterns of the pretrained model and prevent overfitting. Training performance was controlled using accuracy with a threshold of 0.5 as a metric per tile, and the used loss function was binary cross-entropy loss. Model parameters were saved every epoch and the weights of the epoch with the best results were used for evaluation. We further evaluated model performance with five-fold cross-validation, without having any training- and validation-slide and patient overlap. After the training, predictions on the five validation sets were calculated with the corresponding model based on the combined predictions of all tiles of a case. The prediction for a case was calculated using majority voting for the pituitary gland adenoma classifier and the arithmetic mean of the raw predictions (between 0.0 and 1.0) of all the case's tiles for the DNET-GG classifier. These results were then combined and used to calculate true and false-positive rates, which were then used to plot Receiver Operating Characteristic curves, true/false positive frequency histograms, and in conjunction with false-negative rates to plot precision-recall curves.

Since silent corticotroph adenomas only made up 9.7% of the dataset, we decided to train a second neural net on an undersampled training set. The original training set (80% of the complete dataset) consisted of 226,422 tiles of which 59% were positive for LH, 62% for FSH, 22% for ACTH, and 9.4 % were silent corticotroph adenomas. After the undersampling procedure, 54,713 tiles were left of which 43% were positive for LH, 43% for FSH, 43% for ACTH, and 39 % were silent corticotroph adenomas. We assured that at least 30 tiles per WSI were left after undersampling. Again, we used the resnext101_32x8d architecture. The head was trained for five epochs with a maximum learning rate of 10^{-3}. The complete model was then trained for ten epochs with maximum discriminative learning rates ranging from 10^{-7} to 10^{-5}. In both cases, the one-cycle learning rate policy was used with minimum learning rates of $1/25$ of the maximum learning rates.

2.10. Hardware

We implemented our approach on a local server running Ubuntu (18.04 LTS) with one NVIDIA GeForce GTX 1080Ti and one NVIDIA Titan XP, AMD CPU (AMD Ryzen Threadripper 1950X 16 × 3.40 GHz), 128 Gb RAM, CUDA 10.2, and cuDNN 7.

2.11. Availability and Implementation

The datasets generated and analyzed during the presented study are not publicly available, but parts of the pipeline used in this project including training and visualization are available on our Project Homepage.

https://github.com/FAU-DLM/wsi_processing_pipeline, accessed on 15 December 2021.

https://github.com/ChristophNeuner/pituitary_gland_adenomas, accessed on 15 December 2021.

https://github.com/ChristophNeuner/DNET_vs_Ganglioglioma, accessed on 15 December 2021.

3. Results

3.1. Use Case 1: DNET-GG Classifier

We evaluated the performance on the validation set, which made up 20% of the whole dataset and was not used for training. It consisted of 24 slides of ganglioglioma and seven slides of DNET. In total, 29,333 tiles were extracted from the GG slides and 6597 tiles were extracted from the DNET slides for evaluation. No hyperparameter tweaking was performed, which could have led to overfitting on the validation set. On a tile level, the accuracy was 0.936 and on a slide level 0.968. The Brier score on the tile level was 0.053 and 0.022 on the slide level. The AUC on the tile level was 0.93 and 1.00 on the slide level for the ROC curve. The average precision calculated from precision and recall was 0.88 for DNET and 0.97 for GG on the tile level. On the slide level, it was 1.00 for DNET and GG. (Figures 7 and 8)

Model calibration was also evaluated on tile level (Figure 9). We observed tiles that were overconfidently classified by the model as DNET but were in fact GG. DNETs typically contain mucus and have a loosened-up structure. Tiles from GGs which were wrongly classified as DNETs also had a loosened-up structure, which was only artificial.

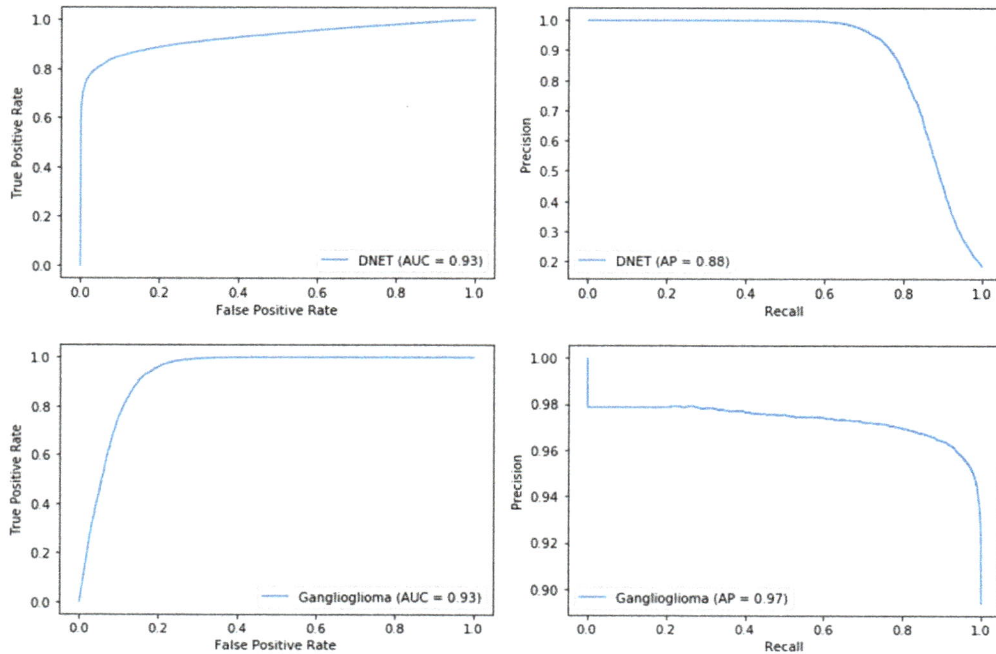

Figure 7. Results | ROC (**left**) and precision recall curves (**right**) on tile level.

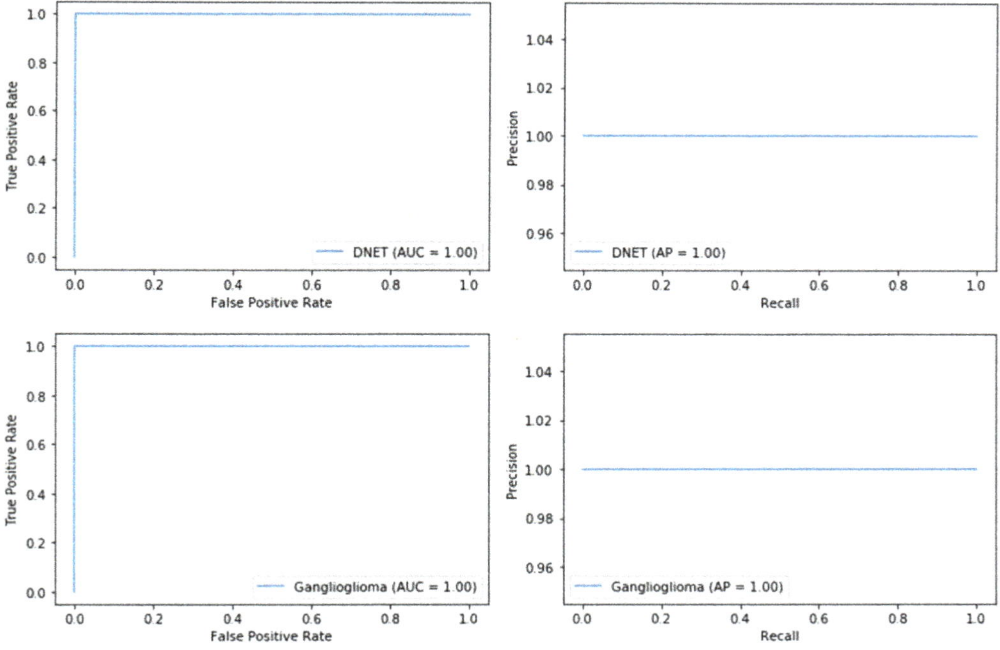

Figure 8. Results | ROC (**left**) and precision recall curves (**right**) on slide level.

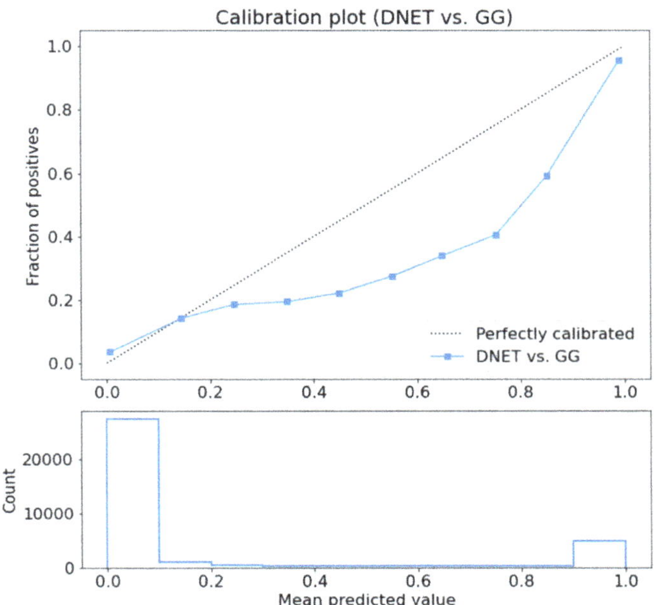

Figure 9. Results | Calibration plot.

3.2. Use Case 2: Pituitary Adenoma Classifier

All CNN were trained to classify the ROIs containing adenoma and surrounding tissue. First, we performed a study to determine which model to use for our classification task. We tested ResNet50, ResNet101, ResNet152, DenseNet121, Xception, Inceptionv4, se_ResNext101_32x4d and ResNext101_32x8d. We compared those models on a predefined validation set with accuracy calculated on a case basis for each class with a threshold of 0.5 (Supplement S5). Inceptionv4, se_ResNext101_32x4d and ResNext101_32x8d showed similar promising results. We decided upon ResNext101_32x8d because of the slightly better test-set results. During training validation, accuracies mostly stayed above training accuracies, and validation loss stayed below training loss values, indicating little to no overfitting on the training dataset. We finally evaluated our model via five-fold cross-validation. For each model within the process of cross-validation, we took 80% of the dataset as training data and 20% as validation data. There was no overlap between these five validation sets. All five validation sets showed similar AUCs with no significant outliers (Supplement S6). Then predictions were made for all tiles of the five validation sets with the respectively corresponding model that was not trained on that particular validation set. Via majority voting with a threshold of 0.5, we then calculated the labels on a case basis and computed AUCs of ROC curves for each class. If more than 50% of the tiles of one case were labeled with the class ACTH, the whole case received the label ACTH.

For ACTH the Brier score was 0.054, for silent ACTH 0.046, for LH 0.069, and for FSH 0.10.

For ACTH the AUC of the ROC curve was 0.97 with a proportion of 44.7% of all cases. The AUC for silent ACTH was 0.86 with a proportion of 9.7%. The AUC for gonadotropic (LH and/or FSH) was 0.98 with a proportion of 55.3%. The AUCs of LH and FSH alone were 0.96 and 0.93 with proportions of 48.1% and 43.8% (Figure 4). Since the silent ACTH cases only made up 9.7% of the dataset, the AUC of 86% of the ROC curve could have simply been a result of guessing. Therefore, we also calculated a precision-recall curve (Figure 10), which resulted in an AUC of 0.71, and, furthermore, trained another neural net on an undersampled dataset as described in the last paragraph of "Training and Evaluation". We reached an accuracy of 88.6% and an AUC of 0.83 for the ROC curve on the validation set for the silent ACTH class (Figure 11).

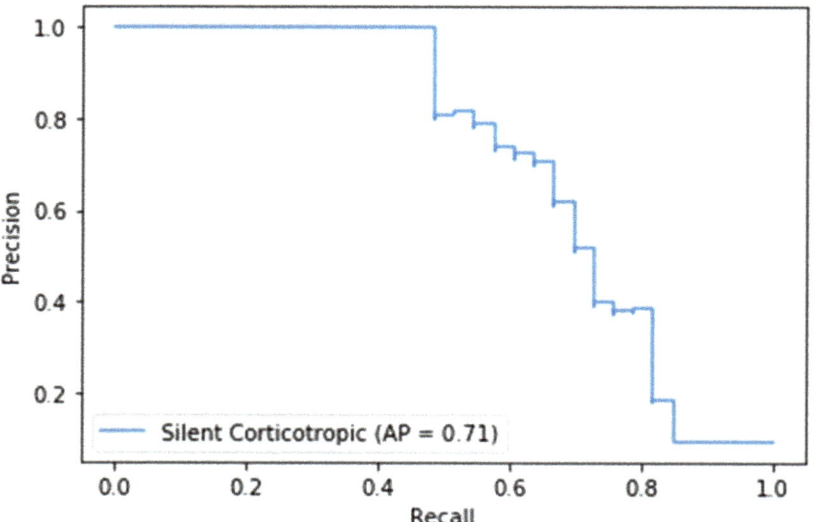

Figure 10. Results | Precision-recall curve for the class silent corticotropic adenoma of the models from the 5-fold cross-validation, which were trained on the unevenly distributed training set, in which silent-corticotroph adenoma made up only 9.7% of the tiles.

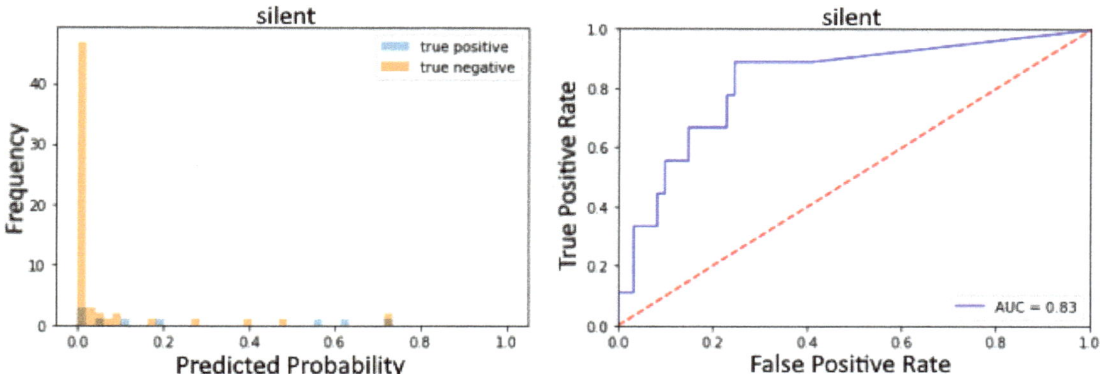

Figure 11. Results | Probability Histogram and ROC-Curve for the class silent corticotroph adenoma of the model that was trained on an undersampled training set in which all four classes were evenly distributed.

We also evaluated the calibration state of our model for the four different classes on slide level (Figure 12). We identified WSI for which the model's prediction differed the most from the true label. Tile quantity and tissue quality had the most influence on the quality of the prediction. If there was only little amount of adenoma present and this tissue was infused with non-pituitary cells, such as blood, connective tissue, or bone, the model had problems predicting the correct class.

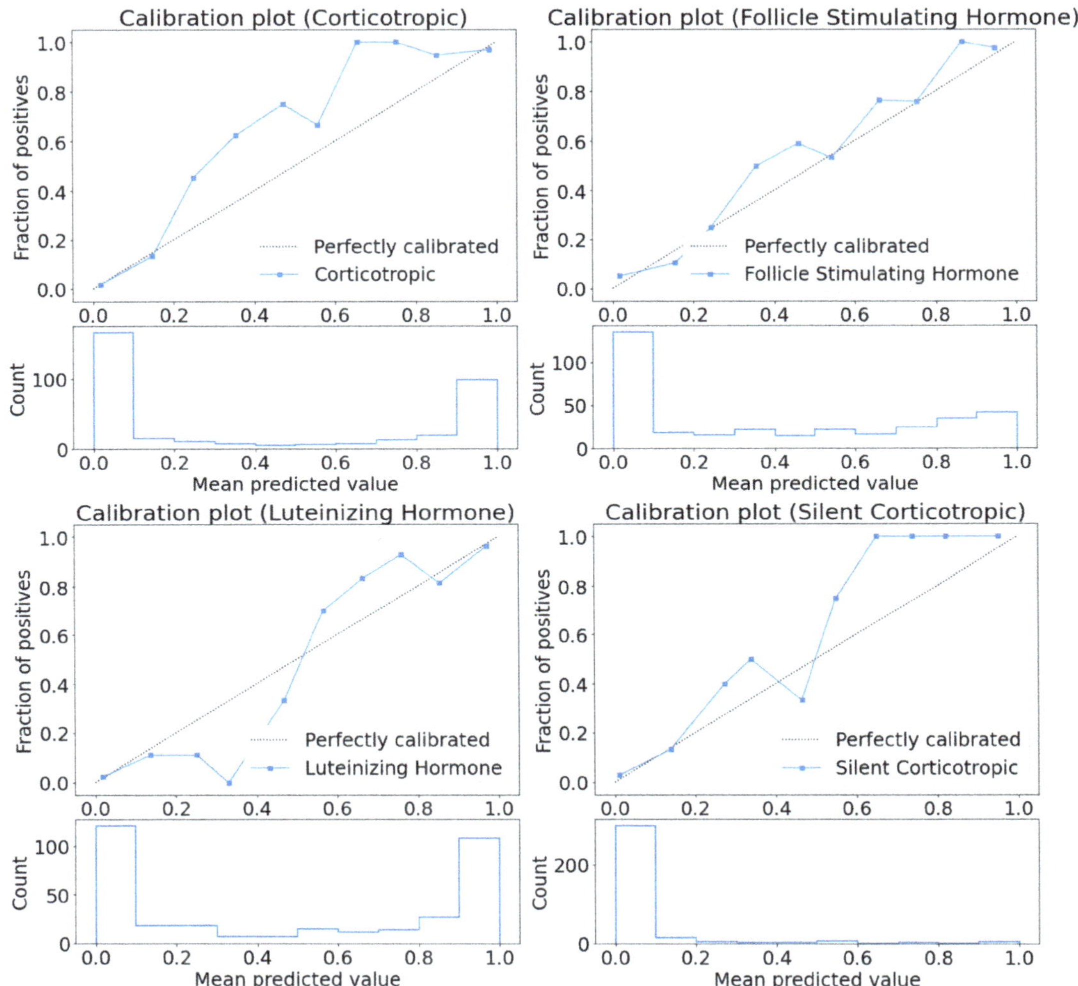

Figure 12. Results | Calibration plot.

4. Discussion

We developed a whole slide image processing library [23] addressing the needs of researchers to assess different DL tasks without the hurdles of complex dataset management. The large size of WSI and annotation of multiple regions of interest tend to increase such technical obstacles. It is also desirable to extract all tiles on the fly during training and only save their spatial information but not the images. This pipeline has the advantage of being more flexible. It is not necessary anymore to repeatedly store extracted tiles as images to disc, saving space and time. Moreover, the evaluation of the trained model requires more steps when dealing with WSI. Results on the tile level are only of limited significance. They have to be transformed into predictions for the complete WSI and the entire case. For histopathologists or expert clinicians addressing a clinical hypothesis, these hurdles may become a real burden. Further, DL experts familiar with the usage of DL frameworks may underestimate the specific handling of digital pathology-associated tasks. The new library provides convenient ways of dealing with WSI in the realm of Neuropathology, thereby facilitating access to DL for both groups of researchers.

Access from and to different levels of magnification, region of interest definition, and handling, as well as dataset splitting, are essential mechanisms and tend to be technically intricate. The library manages these crucial steps and offers default parameters enabling the user to focus on the problem-specific tasks. For the specific use-cases addressed in this study, the library facilitated the management of pre-extracted image patches for a given patient as well as extraction of image patches on the fly from predefined ROI. Our evaluation of different state-of-the-art model architectures to identify the most suitable model for the problem-specific tasks, i.e., best classification results and least overfitting, resulted in the selection of resnet50 for the first use-case and the resnext101_32x8d [35,37] architecture for the second use-case. We believe that these rather large networks with lots of parameters worked well, because of their large input image size of 512×512 pixels. On smaller images, networks with fewer parameters tend to work better in our experience [3]. A crucial step in our pipeline was the way of image preprocessing. One part of this aspect was image augmentation to increase the variance presented to the network [42]. Normalization of the input data was performed with the mean and standard deviation of our own dataset. Fastai [30] does this conveniently for the user.

Use-case 1: In the first use-case, we developed a DL approach to distinguish between two epilepsy-associated tumors, the GG and the DNET. Since unlike DNET, some GG can undergo malignant transformation [11,12], a precise distinction between these two entities is crucial. We were able to demonstrate that a CNN can differentiate between these two entities with a very high accuracy only using H&E-stained slides. This confirms the potential of DL in assisting pathologists in their decision-making diagnostic process and to eventually reducing the necessity for further stains.

Use-case 2: In the second use-case addressed, we developed a DL approach to help to diagnose the entity of pituitary adenomas without the necessity of additional immunohistochemical stainings. Additionally, we could show that even a clinical parameter, such as the clinical occurrence of M. Cushing of corticotroph adenomas, might be hidden within the tissue; however, it could successfully be recognized by our neural network approach. This evidence supports the hypothesis that clinical parameters can be found within histomorphology, and that distinct features may be revealed by DL in terms of imaging biomarkers. Guided Grad-CAMs [32] could now be used to visualize the decision making and to teach pathologists which morphological structures are crucial for the network in its decision-making process.

We addressed the classification task on predictions per tile and collected all votes for the given slides of a patient's case. We then obtained the final diagnosis by majority voting to obtain predictions on a case basis. If more than 50% of the tiles of one case were labeled with one class, the case was given that class label. We chose that option for two reasons.

First, different from finding metastasis in lymph nodes where high sensitivity is needed, histological slides from pituitary adenomas usually contain massive adenoma; hence, most of the tissue on the slide belongs to the tumor. Second, time was not a major concern. We could simply take and analyze all possible tiles instead of only taking a representative batch for inference.

Limitations and Potential Solutions Moving into the Future

A well-recognized obstacle in digital pathology represents batch effects including variation in staining intensity or fixation artifacts [4,43]. We contained such batch effects in our input data through hand-picked ROI and normalization. We did not directly address the problem of stain normalization [44]for this dataset, because all staining was performed in a single lab, and only one device was used for scanning. For further usage of our model in a production environment with whole slide images from other institutes, this would be crucial. We are continuously working on this issue to make our models more robust in the future.

Histopathology analysis represents a gold standard in tumor diagnosis as it often directs further treatment. Adenomas of the pituitary gland, although routinely classified by immunohistochemical profiling of their neuroendocrine axis, are in urgent need of a clinically meaningful histopathology classification of their risk for relapse. This was partially addressed by the WHO classification from 2004 and 2016. The criteria of atypia to label more aggressive adenomas has been removed, however, as it has not proved a predictive marker [17,45]. The "silent" corticotroph class of our dataset did represent another clinical parameter of interest and was remarkably well recognized by our network, even in the evenly distributed dataset. The good classification result of the "silent" corticotroph class in our study shows that neuronal networks are capable of revealing such clinical information hidden within tissue slides and, hence, it may also be possible to extract a clinical relapse parameter from tissue slides via DL. However, due to the lack of datasets stained at different labs, digitized from different scanners, and the size of the dataset, our well-performing models may be unsuitable for clinical practice yet.

In conclusion, we developed a convenient open-access library compatible with fastai to support hypothesis-driven DL research projects in the realm of neuropathology.

It helps in managing the dataset by assigning hierarchy levels such as patients, cases, and slides, thereby making it easily possible to split the dataset for training and evaluation. The library consists of building blocks fully compatible with fastai for easy integration and usage of the full spectrum of fastai functionality. Additionally, many visualization methods for evaluation are implemented.

Both use-cases demonstrated the successful diagnosis of adenoma of the pituitary gland and distinguishing between DNET and GG by H&E-stained slides only and without the necessity of cost- and labor-intense immunohistochemistry staining.

Supplementary Materials: The following are available online at https://www.mdpi.com/article/10.3390/app12010013/s1, Supplement S1: Dataset, Supplement S2: Class Distribution, Supplement S3: Custom head (Pytorch), Supplement S4: Learning rate finder pituitary adenoma classifier, Supplement S5: Evaluated Networks, Supplement S6: AUCs of the ROC-curves for the five validation sets of 5-fold cross-validation, Supplement S7: QuPath, Supplement S8: ROIs with overlaid grids, Supplement S9: Tissue filtering.

Author Contributions: Conceptualization, C.N., S.J., R.C., A.W. and I.B.; methodology, C.N. and S.J.; software, C.N.; validation, C.N., S.J., R.C. and I.B.; formal analysis, C.N. and S.J.; investigation, C.N. and S.J.; resources, S.J., I.B., S.M.S. and M.B.; data curation, C.N. and A.P.; writing—original draft preparation, C.N.; writing—review and editing, S.J., R.C., I.B., S.M.S., M.B. and A.W.; visualization, C.N.; supervision, S.J.; project administration, S.J.; funding acquisition, S.J. and I.B. All authors have read and agreed to the published version of the manuscript.

Funding: This research was funded by the Interdisciplinary Center for Clinical Research (IZKF) at the University Hospital of the University of Erlangen-Nuremberg, grant number Junior Project J81.

Institutional Review Board Statement: Not applicable.

Informed Consent Statement: Not applicable.

Data Availability Statement: The code used here is available from the following three GitHub repositories: https://github.com/FAU-DLM/wsi_processing_pipeline, accessed on 15 December 2021, https://github.com/ChristophNeuner/pituitary_gland_adenomas, accessed on 15 December 2021, https://github.com/ChristophNeuner/DNET_vs_Ganglioglioma, accessed on 15 December 2021. The whole-slide images used here are not publicly available.

Acknowledgments: The present work was performed in fulfillment of the requirements of the Friedrich-Alexander Universität Erlangen-Nürnberg (FAU) for obtaining the degree 'Dr. med.' of Christoph Neuner. The work was supported by the Interdisciplinary Center for Clinical Research (IZKF) at the University Hospital of the University of Erlangen-Nuremberg (Junior Project "J81"). We would also like to thank NVIDIA for the donation of a Titan XP.

Conflicts of Interest: The authors declare no conflict of interest.

References

1. Bejnordi, B.E.; Veta, M.; Van Diest, P.J.; Van Ginneken, B.; Karssemeijer, N.; Litjens, G.; Van Der Laak, J.A.W.M.; Hermsen, M.; Manson, Q.F.; Balkenhol, M.; et al. Diagnostic Assessment of Deep Learning Algorithms for Detection of Lymph Node Metastases in Women With Breast Cancer. *JAMA* **2017**, *318*, 2199–2210. [CrossRef]
2. Arvaniti, E.; Fricker, K.S.; Moret, M.; Rupp, N.; Hermanns, T.; Fankhauser, C.; Wey, N.; Wild, P.J.; Rüschoff, J.H.; Claassen, M. Automated Gleason grading of prostate cancer tissue microarrays via deep learning. *Sci. Rep.* **2018**, *8*, 12054. [CrossRef]
3. Kubach, J.; Muhlebner-Fahrngruber, A.; Soylemezoglu, F.; Miyata, H.; Niehusmann, P.; Honavar, M.; Rogerio, F.; Kim, S.H.; Aronica, E.; Garbelli, R.; et al. Same same but different: A Web-based deep learning application revealed classifying features for the histopathologic distinction of cortical malformations. *Epilepsia* **2020**, *61*, 421–432. [CrossRef]
4. Tang, Z.; Chuang, K.V.; DeCarli, C.; Jin, L.-W.; Beckett, L.; Keiser, M.J.; Dugger, B.N. Interpretable classification of Alzheimer's disease pathologies with a convolutional neural network pipeline. *Nat. Commun.* **2019**, *10*, 2173. [CrossRef]
5. Van der Laak, J.; Litjens, G.; Ciompi, F. Deep learning in histopathology: The path to the clinic. *Nat. Med.* **2021**, *27*, 775–784. [CrossRef]
6. Esteva, A.; Kuprel, B.; Novoa, R.A.; Ko, J.; Swetter, S.M.; Blau, H.M.; Thrun, S. Dermatologist-level classification of skin cancer with deep neural networks. *Nature* **2017**, *542*, 115–118. [CrossRef]
7. Janowczyk, A.; Madabhushi, A. Deep learning for digital pathology image analysis: A comprehensive tutorial with selected use cases. *J. Pathol. Inform.* **2016**, *7*, 29. [CrossRef]
8. Coudray, N.; Ocampo, P.S.; Sakellaropoulos, T.; Narula, N.; Snuderl, M.; Fenyö, D.; Moreira, A.L.; Razavian, N.; Tsirigos, A. Classification and mutation prediction from non–small cell lung cancer histopathology images using deep learning. *Nat. Med.* **2018**, *24*, 1559–1567. [CrossRef]
9. Steiner, D.F.; Macdonald, R.; Liu, Y.; Truszkowski, P.; Hipp, J.D.; Gammage, C.; Thng, F.; Peng, L.; Stumpe, M.C. Impact of Deep Learning Assistance on the Histopathologic Review of Lymph Nodes for Metastatic Breast Cancer. *Am. J. Surg. Pathol.* **2018**, *42*, 1636–1646. [CrossRef]
10. Blümcke, I.; Coras, R.; Wefers, A.K.; Capper, D.; Aronica, E.; Becker, A.; Honavar, M.; Stone, T.J.; Jacques, T.S.; Miyata, H.; et al. Review: Challenges in the histopathological classification of ganglioglioma and DNT: Microscopic agreement studies and a preliminary genotype-phenotype analysis. *Neuropathol. Appl. Neurobiol.* **2018**, *45*, 95–107. [CrossRef]
11. Majores, M.; von Lehe, M.; Fassunke, J.; Schramm, J.; Becker, A.J.; Simon, M. Tumor recurrence and malignant progression of gangliogliomas. *Cancer* **2008**, *113*, 3355–3363. [CrossRef] [PubMed]
12. Selvanathan, S.K.; Hammouche, S.; Salminen, H.J.; Jenkinson, M. Outcome and prognostic features in anaplastic ganglioglioma: Analysis of cases from the SEER database. *J. Neuro-Oncol.* **2011**, *105*, 539–545. [CrossRef]
13. Thom, M.; Toma, A.; An, S.; Martinian, L.; Hadjivassiliou, G.; Ratilal, B.; Dean, A.; McEvoy, A.; Sisodiya, S.M.; Brandner, S. One Hundred and One Dysembryoplastic Neuroepithelial Tumors: An Adult Epilepsy Series With Immunohistochemical, Molecular Genetic, and Clinical Correlations and a Review of the Literature. *J. Neuropathol. Exp. Neurol.* **2011**, *70*, 859–878. [CrossRef] [PubMed]
14. Slegers, R.J.; Blumcke, I. Low-grade developmental and epilepsy associated brain tumors: A critical update 2020. *Acta Neuropathol. Commun.* **2020**, *8*, 27. [CrossRef] [PubMed]
15. Ezzat, S.; Asa, S.L.; Couldwell, W.T.; Barr, C.E.; Dodge, W.E.; Vance, M.L.; McCutcheon, I.E. The prevalence of pituitary adenomas. *Cancer* **2004**, *101*, 613–619. [CrossRef]
16. Aflorei, E.D.; Korbonits, M. Epidemiology and etiopathogenesis of pituitary adenomas. *J. Neuro-Oncol.* **2014**, *117*, 379–394. [CrossRef]
17. Inoshita, N.; Nishioka, H. The 2017 WHO classification of pituitary adenoma: Overview and comments. *Brain Tumor Pathol.* **2018**, *35*, 51–56. [CrossRef]
18. Vizcarra, J.C.; Gearing, M.; Keiser, M.J.; Glass, J.D.; Dugger, B.N.; Gutman, D.A. Validation of machine learning models to detect amyloid pathologies across institutions. *Acta Neuropathol. Commun.* **2020**, *8*, 59. [CrossRef]
19. Signaevsky, M.; Prastawa, M.; Farrell, K.; Tabish, N.; Baldwin, E.; Han, N.; Iida, M.A.; Koll, J.; Bryce, C.; Purohit, D.; et al. Artificial intelligence in neuropathology: Deep learning-based assessment of tauopathy. *Lab. Investig.* **2019**, *99*, 1019–1029. [CrossRef]
20. Koga, S.; Ikeda, A.; Dickson, D.W. Deep learning-based model for diagnosing Alzheimer's disease and tauopathies. *Neuropathol. Appl. Neurobiol.* **2021**. [CrossRef]
21. Vega, A.R.; Chkheidze, R.; Jarmale, V.; Shang, P.; Foong, C.; Diamond, M.I.; White, C.L.; Rajaram, S. Deep learning reveals disease-specific signatures of white matter pathology in tauopathies. *Acta Neuropathol. Commun.* **2021**, *9*, 170. [CrossRef] [PubMed]
22. Jin, L.; Shi, F.; Chun, Q.; Chen, H.; Ma, Y.; Wu, S.; Hameed, N.U.F.; Mei, C.; Lu, J.; Zhang, J.; et al. Artificial intelligence neuropathologist for glioma classification using deep learning on hematoxylin and eosin stained slide images and molecular markers. *Neuro-Oncology* **2020**, *23*, 44–52. [CrossRef]
23. Neuner, C. Python-Wsi-Preprocessing. GitHub. 2019. Available online: https://github.com/FAU-DLM/python-wsi-preprocessing (accessed on 16 December 2021).
24. Veta, M.; Heng, Y.J.; Stathonikos, N.; Bejnordi, B.E.; Beca, F.; Wollmann, T.; Rohr, K.; Shah, M.A.; Wang, D.; Rousson, M.; et al. Predicting breast tumor proliferation from whole-slide images: The TUPAC16 challenge. *Med. Image Anal.* **2019**, *54*, 111–121. [CrossRef] [PubMed]

25. Eriksson, D. Python-Wsi-Preprocessing. GitHub. 2018. Available online: https://github.com/deroneriksson/python-wsi-preprocessing (accessed on 28 December 2019).
26. Bankhead, P.; Loughrey, M.B.; Fernández, J.A.; Dombrowski, Y.; McArt, D.G.; Dunne, P.D.; McQuaid, S.; Gray, R.T.; Murray, L.J.; Coleman, H.G.; et al. QuPath: Open source software for digital pathology image analysis. *Sci. Rep.* **2017**, *7*, 16878. [CrossRef] [PubMed]
27. Paeng, K.; Hwang, S.; Park, S.; Kim, M. A Unified Framework for Tumor Proliferation Score Prediction in Breast Histopathology. In *Deep Learning in Medical Image Analysis and Multimodal Learning for Clinical Decision Support*; Springer: Cham, Switzerland, 2017; pp. 231–239. [CrossRef]
28. Pascale, D. A Reviw of RGB Color Spaces. 6 October 2003. Available online: https://www.babelcolor.com/index_htm_files/A%20review%20of%20RGB%20color%20spaces.pdf (accessed on 28 November 2021).
29. Zenil, H. HSV Colors. 1 March 2011. Available online: https://demonstrations.wolfram.com/HSVColors/ (accessed on 28 November 2021).
30. Howard, J.; Gugger, S. Fastai: A Layered API for Deep Learning. *Information* **2020**, *11*, 108. [CrossRef]
31. Paszke, A.; Gross, S.; Massa, F.; Lerer, A.; Bradbury, J.; Chanan, G.; Killeen, T.; Lin, Z.; Gimelshein, N.; Antiga, L.; et al. PyTorch: An Imperative Style, High-Performance Deep Learning Library. *Adv. Neural Inf. Process. Syst.* **2019**, *32*. Available online: https://proceedings.neurips.cc/paper/2019/file/bdbca288fee7f92f2bfa9f7012727740-Paper.pdf (accessed on 16 December 2021).
32. Selvaraju, R.R.; Das, A.; Vedantam, R.; Cogswell, M.; Parikh, D.; Batra, D. Grad-CAM: Why did you say that? *arXiv* **2016**, arXiv:1611.07450.
33. Simonyan, K.; Zisserman, A. Very Deep Convolutional Networks for Large-Scale Image Recognition. *arXiv* **2014**, arXiv:1409.1556.
34. Deng, J.; Dong, W.; Socher, R.; Li, L.-J.; Li, K.; Li, F.-F. ImageNet: A large-scale hierarchical image database. In Proceedings of the 2009 IEEE Computer Society Conference on Computer Vision and Pattern Recognition, Miami, FL, USA, 20–25 June 2009; pp. 248–255.
35. Saining, X.; Ross, G.; Piotr, D.; Zhuowen, T.; He, K. Aggregated Residual Transformations for Deep Neural Networks. *arXiv* **2016**, arXiv:1611.05431.
36. Cadene, R. Pretrained PyTorch Models. GitHub. 2019. Available online: https://github.com/Cadene/pretrained-models.pytorch (accessed on 16 December 2021).
37. Wu, R.; Yan, S.; Shan, Y.; Dang, Q.; Sun, G. Deep Image: Scaling up Image Recognition. *arXiv* **2015**, arXiv:1501.02876.
38. Wong, S.C.; Gatt, A.; Stamatescu, V.; McDonnell, M.D. Understanding Data Augmentation for Classification: When to Warp? *arXiv* **2016**, arXiv:1609.08764.
39. Micikevicius, P.; Narang, S.; Alben, J.; Diamos, G.; Elsen, E.; Garcia, D.; Ginsburg, B.; Houston, M.; Kuchaiev, O.; Venkatesh, G.; et al. Mixed Precision Training. *arXiv* **2017**, arXiv:1710.03740.
40. Kingma, D.P.; Ba, J. Adam: A Method for Stochastic Optimization. *arXiv* **2014**, arXiv:1412.6980.
41. Smith, L.N. Cyclical learning rates for training neural networks. *arXiv* **2015**, arXiv:1506.01186.
42. Perez, L.; Wang, J. The Effectiveness of Data Augmentation in Image Classification using Deep Learning. *arxiv* **2017**, arXiv:1712.04621.
43. Madabhushi, A.; Lee, G. Image analysis and machine learning in digital pathology: Challenges and opportunities. *Med. Image Anal.* **2016**, *33*, 170–175. [CrossRef]
44. Anghel, A.; Stanisavljevic, M.; Andani, S.; Papandreou, N.; Rüschoff, J.H.; Wild, P.; Gabrani, M.; Pozidis, H. A High-Performance System for Robust Stain Normalization of Whole-Slide Images in Histopathology. *Front. Med.* **2019**, *6*, 193. [CrossRef] [PubMed]
45. Mete, O.; Lopes, M.B. Overview of the 2017 WHO Classification of Pituitary Tumors. *Endocr. Pathol.* **2017**, *28*, 228–243. [CrossRef]

Article

Automated Detection of Gastric Cancer by Retrospective Endoscopic Image Dataset Using U-Net R-CNN

Atsushi Teramoto [1,*,†], Tomoyuki Shibata [2,†], Hyuga Yamada [2], Yoshiki Hirooka [2], Kuniaki Saito [1] and Hiroshi Fujita [3]

1   School of Medical Sciences, Fujita Health University, Toyoake 470-1192, Japan; saitok@fujita-hu.ac.jp
2   Department of Gastroenterology and Hepatology, Fujita Health University, Toyoake 470-1192, Japan; shibat03@fujita-hu.ac.jp (T.S.); hyugayama1988@yahoo.co.jp (H.Y.); yoshiki.hirooka@fujita-hu.ac.jp (Y.H.)
3   Faculty of Engineering, Gifu University, Gifu 501-1194, Japan; fujita@fjt.info.gifu-u.ac.jp
*   Correspondence: teramoto@fujita-hu.ac.jp
†   These authors contributed equally to this work.

Abstract: Upper gastrointestinal endoscopy is widely performed to detect early gastric cancers. As an automated detection method for early gastric cancer from endoscopic images, a method involving an object detection model, which is a deep learning technique, was proposed. However, there were challenges regarding the reduction in false positives in the detected results. In this study, we proposed a novel object detection model, U-Net R-CNN, based on a semantic segmentation technique that extracts target objects by performing a local analysis of the images. U-Net was introduced as a semantic segmentation method to detect early candidates for gastric cancer. These candidates were classified as gastric cancer cases or false positives based on box classification using a convolutional neural network. In the experiments, the detection performance was evaluated via the 5-fold cross-validation method using 1208 images of healthy subjects and 533 images of gastric cancer patients. When DenseNet169 was used as the convolutional neural network for box classification, the detection sensitivity and the number of false positives evaluated on a lesion basis were 98% and 0.01 per image, respectively, which improved the detection performance compared to the previous method. These results indicate that the proposed method will be useful for the automated detection of early gastric cancer from endoscopic images.

Keywords: gastric cancer; endoscopy; deep learning; convolutional neural network

Citation: Teramoto, A.; Shibata, T.; Yamada, H.; Hirooka, Y.; Saito, K.; Fujita, H. Automated Detection of Gastric Cancer by Retrospective Endoscopic Image Dataset Using U-Net R-CNN. *Appl. Sci.* **2021**, *11*, 11275. https://doi.org/10.3390/app112311275

Academic Editor: Fabio La Foresta

Received: 3 November 2021
Accepted: 26 November 2021
Published: 28 November 2021

1. Introduction

1.1. Background

Gastric cancer (GC) is one of the most common malignant tumors of the stomach mucosa. According to global statistics, GC is the second leading cause of cancer deaths, and the number of patients with GC is increasing due to changes in dietary habits and longer life expectancy [1,2]. GC may be effectively treated if detected early. Therefore, early detection and treatment of gastric cancer are essential.

Radiography and endoscopy are used to screen for GC. During endoscopy, a specialist inserts an endoscope through the patient's mouth or nose and directly observes the mucous membrane of the digestive tract to detect abnormalities. The detection sensitivity of GC by endoscopy is high, and if lesions are found during the examination then tissue may be collected and simple treatment can be given [3].

However, specialists who perform the procedure need to detect abnormalities while operating the endoscope, making the examination process sophisticated and complicated. This results in widely varying diagnostic accuracy, and some studies have reported that lesions were missed in 22.2% of cases [4]. If physicians could use the results of computerized image analysis and detect abnormalities during the examination, they could solve some of these problems and detect GC at an early stage. Deep learning, an artificial intelligence technology, has recently been confirmed to have high capability for image recognition by several studies that were conducted in the medical field [5–10]. Therefore, we focused on the automated detection of GC in endoscopic images using a computer-aided diagnostic method based on deep learning technology.

1.2. Related Works

There are many studies on deep learning for the diagnosis of GC using endoscopic images, including studies on the classifications between GC and healthy subjects and the automated recognition of GC regions.

Shichijo et al. investigated the prediction of Helicobacter pylori infection using a convolutional neural network (CNN) and obtained a sensitivity of 88.9% and specificity of 87.4% [11]. Li et al. developed a method to discriminate between GC and normal tissue using magnified narrow-band imaging (NBI) [12]. They used Inception-v3 as the CNN model for classification and obtained a sensitivity of 91.18% and specificity of 90.64%. Zhang et al. developed a method to classify precancerous diseases (polyp, ulcer, and erosion) using CNN and obtained a classification accuracy of 88.9% [13].

Hirasawa et al. developed a single-shot multi-box detector (SSD), an object detection model, for the automated detection of early-stage GC [14]. The sensitivity of detection was 92.2% and the positive predictive value was 30.6%. Sakai et al. also developed a method for object detection of GC by classifying GC regions and normal regions using micropatch endoscopic images [15]. The detection sensitivity and specificity of the method were 80.0% and 94.8%, respectively.

We proposed a method for extracting the presence and invasive regions of early GC using Mask R-CNN, which can perform both object detection and segmentation [16]. We showed that the automated detection sensitivity for early GC was 96.0% and that the segmentation concordance was 71%. Although the method had sufficient detection sensitivity, the average number of false positives (FP) was 0.10 per image (3.0 per patient). The Mask R-CNN used in this study introduced an object detection model for common natural images. It captured the clear contour of the object in the image, so that lesions with a relatively clear shape that caused unevenness were detected correctly. On the other hand, many lesions of early GC, in which only the surface of the gastric mucosa was cancerous, were not detected correctly by the object detection model because contours were unclear.

A CNN used for segmentation rather than object detection analyzes patterns in the local regions of the image and divides the entire image into regions by determining whether they match the patterns to be extracted. This behavior involved in determining individual regions while observing details is similar to that used by an expert physician when observing the gastric cavity, and segmentation techniques may be able to improve the accuracy of automated lesion detection. On the other hand, many small regions are often observed in the segmentation output by CNNs. Excluding these using the FP reduction technique may greatly reduce the number of FPs and improve detection performance. Therefore, segmentation techniques that exclude small excess regions are effective for the automated detection of GC and identification of the extent of invasion.

1.3. Objective

In this study, we develop a deep learning model that may accurately detect the presence of GC and its extent of invasion using endoscopic images. We propose a novel deep learning model, U-Net R-CNN, which is a combination of the U-Net segmentation process and CNN for image classification to eliminate FPs. The efficacy of this method is confirmed using endoscopic images of early GC and healthy subjects.

2. Methods

2.1. Outline

The outline of the GC detection by U-Net R-CNN proposed in this study is shown in Figure 1. The endoscopic images were given to the U-Net, the initial candidate regions of the GC were extracted, and the bounding box of each extracted region was obtained. The obtained image patterns in the bounding boxes were subjected to a CNN to classify FPs and the final candidates (box classification). The final candidate regions where FPs are excluded were outputs used for diagnosis.

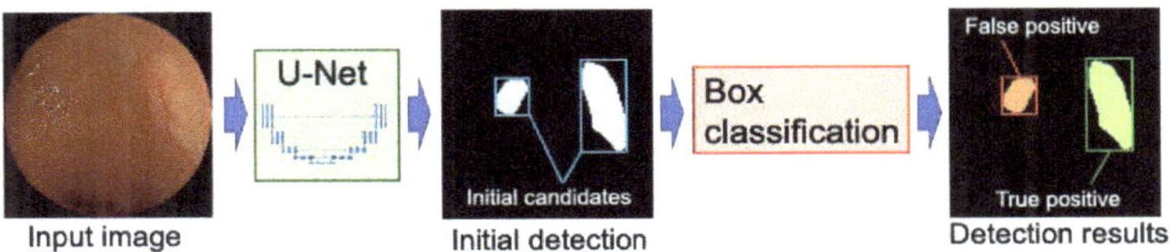

Figure 1. Outline of the proposed method.

2.2. Image Dataset

The image dataset used in this study was same as that used in our previous study. Patient details and other information may be found in our previous study [16]. For this study, 42 healthy and 93 cases (94 lesions) of GC for preoperative examinations were collected between 16 July 2013 and 30 August 2017 at the Fujita Health University Hospital. The numbers of images for the above two categories were 1208 and 533, respectively. The endoscopic images were obtained from multiple directions if a lesion was found during the examination. Table 1 shows the characteristics of gastric cancer patients and lesions [16]. Regarding the healthy subjects, we reassessed the cases endoscopists diagnosed without any abnormalities. When we did not find a specific lesion, such as a tumor, polyps, or gastritis, and a regular arrangement of collecting venules was observed in the mucosa, we considered this as "healthy" [17].

Table 1. Clinical characteristics of gastric cancer lesions in the dataset.

Characteristics	Number of Cases
Tumor position	
Lower third	33
Middle third	52
Upper third	9
Macroscopic classification	
Type 0-I	0
Type 0-IIa	10
Type 0-IIb	0
Type 0-IIc	63
Type 0-III	0
Type 0 mixed (0-IIa + IIc and IIc + III)	21
Depth of tumor invasion	
T1a (mucosa)	71
T1b (submucosa)	23
Histopathological classification	
Undifferentiated	13
Differentiated	75
Mixed	6

We obtained images using upper endoscope units (EG-L600ZW7; Fujifilm Corp., Tokyo, Japan; GIF-290Z, GIF-HQ290, GIF-XP290N, GIF-260Z; Olympus Medical Systems, Co., Ltd., Tokyo, Japan) and standard endoscopic video systems (VP-4450HD/LL-4450, Fujifilm Corp., Tokyo, Japan; and EVIS LUCERA CV-260/CLV-260; EVIS LUCERA ELITE CV-290/CLV-290SL; Olympus Medical Systems, Tokyo, Japan).

All images were captured using standard white light and stored in JPEG file format. The matrix size of the images ranged from 640 × 480 to 1440 × 1080 pixels. To make the matrix size consistent, we resized all images to 512 × 512 pixels. The circular field of view was adopted to avoid differences among the endoscopic instruments and to facilitate data augmentation, before being trimmed as a circle.

The expert endoscopist (TS), who was certified by the Japan Gastroenterological Endoscopy Society, made the ground truth of a label image. To make the label images, we used the original Python software.

2.3. Data Augmentation

Rotation and inversion invariance can be established because the endoscope may capture images of the gastric condition from various angles. Therefore, various images may be created by rotating or flipping each of the collected images. In this study, to ensure stable deep learning performance, we prepared, rotated, and flipped original images for data augmentation and used them for training. Using our in-house software, we generated images by setting the rotation pitch of the images of GC and healthy subjects to 6° and 10°, respectively, so that the numbers of images of GC and healthy subjects were equal [16].

2.4. Initial Detection

We extracted the candidate regions for early GC from endoscopic images. For this task, we employed U-Net, a semantic segmentation technique [18], which was first proposed in 2015 as a method for extracting cell regions in microscopic images and widely used in fields other than medical imaging since. The network structure is shown in Figure 2. U-Net consists of five convolutional and pooling layers, followed by five encoder layers (upscaling layers). When an image is given to the input layer, the encoder layer in the first half extracts the features of the image. Then, the decoder layer in the second half outputs a segmented label image based on the extracted features. In addition, the encoder and decoder layers are connected to each other and the high-resolution information from the encoder layer is delivered directly to the decoder layer on the other side, thereby increasing the resolution of the label image. U-Net provides an initial candidate region for early GC corresponding to the input image. As for the U-Net parameters, the Dice coefficient was used as the loss function (the definition of the Dice index is described in Section 2.6), with the Adam algorithm [19] as the optimization algorithm, 0.0001 as the learning coefficient, 100 as the number of training sessions, and 8 as the batch size.

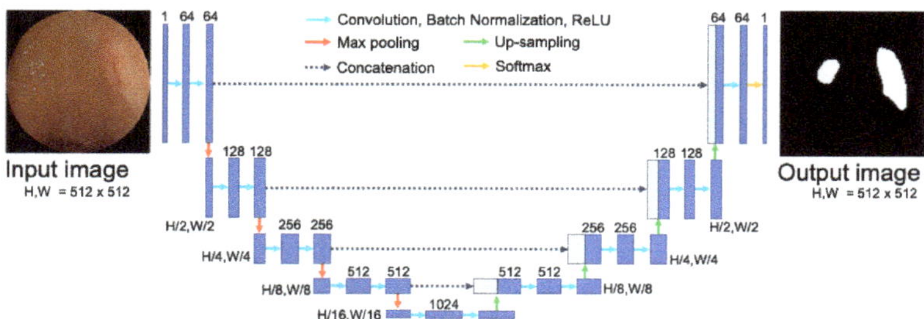

Figure 2. Architecture of U-Net.

2.5. Box Classification

The detected candidate regions included many over-detected regions (FPs). These FPs may be recognized and reduced using a different approach from the segmented U-Net for segmentation. In the box classification part of the proposed U-Net R-CNN, FPs are eliminated from the candidate regions by another CNN (Figure 3).

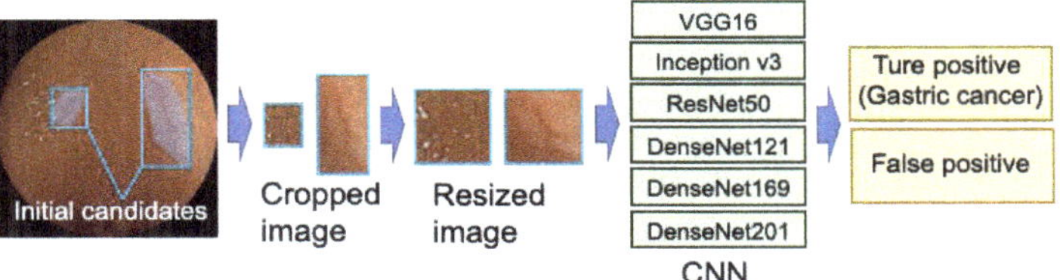

Figure 3. Architecture of the box classification.

First, the input image was provided to U-Net and the output labeled image of U-Net was automatically binarized by Otsu's method [20], followed by the labeling process to pick up individual candidate regions. The bounding box of the candidate region was then cut out and input to the CNN to classify the candidate region as GC or false positive. Finally, the regions that the CNN identified as GC were used as the final detection results.

For the CNN architecture, we introduced VGG-16 [21], Inception v3 [22], and ResNet50 [23], as well as DenseNet121, 169, and 201 [24], then selected the best model by comparing them. These CNN models were pretrained using the ImageNet dataset, which has a much larger number of training image samples than our original dataset. For the classification of GC and FPs, we replaced the fully connected layers of the original CNN models with three layers having 1024, 256, and 2 units.

For the input of the CNN, the image of the candidate region was resized to 224×224 pixels and the optimization algorithm was the Adam algorithm with a learning rate of 0.0001 and 200 training epochs. For data augmentation, vertical and horizontal image flipping were performed randomly.

2.6. Evaluation Metrics

We defined the evaluation metrics to assess the detection and segmentation performance of the proposed method. As for the detection sensitivity, when the GC region obtained by the proposed method and the ground truth region specified by a gastrointestinal specialist overlapped, we evaluated that the target GC was detected correctly. In the endoscopic examination, the same GC region was often observed in a number of images because images were taken from many angles. Among these images, some images may have subtle patterns that are difficult to identify. Therefore, we evaluated the performance of two counting methods. The first method involved simply counting the number of images for which GC was detected correctly (image-based sensitivity), while second method determined whether one lesion was correctly detected in at least one image (lesion-based sensitivity). Regarding the FPs, the total number of detected healthy regions in healthy cases were counted; we calculated FPs per case by dividing them by the total number of healthy subjects.

Although the main task of this study was to detect objects to recognize the presence of GC, U-Net may extract object regions. Therefore, we evaluated the accuracy of region extraction using Dice and Jaccard coefficients.

Di and Ji were calculated using the following formulas to evaluate the similarity between the detected region and the ground truth created by a gastrointestinal specialist:

$$Di = 2\,|\,A \cap B\,|\,/(\,|\,A\,| + |\,B\,|) \times 100\ [\%] \tag{1}$$

$$Ji = |\,A \cap B\,|\,/\,|\,A \cup B\,| \times 100\ [\%] \tag{2}$$

where A and B are two sets. Here, A indicates the ground truth GC region specified by a specialist and B indicates the detected region identified by the proposed method.

Di and Ji were evaluated in the two groups. First, all images containing GC areas were used to evaluate the overall extraction accuracy. In the second method, only the regions detected by the method were evaluated to confirm the accuracy of the invasion area when GC was detected.

In the evaluation, we used a cross-validation method [25]. In this method, the dataset was split into k groups (called k-fold cross-validation). The network was trained using the k-1 subset; the remaining subset was used for the test. By repeating the above process k times, the test results for all data can be obtained. The overall model accuracy can be calculated by summarizing all test results. In our evaluation, five-fold cross-validation (k = 5) was introduced; 137 cases were randomly divided into five groups. Here, the images of the same case were not assigned to both training and test data.

This study was approved by the institutional review board of Fujita Health University, informed consent was obtained from the patients in the form of an opt-out in the endoscopic center of Fujita Health University Hospital, and all data were anonymized (No. HM17-226). This study was conducted in accordance with the World Medical Association Declaration of Helsinki.

The calculations of the initial detection and FP reduction were performed using original Python software using an AMD Ryzen9 3950X processor (16 CPU cores, 4.7 GHz) with 128 GB of DDR4 memory. Trainings of CNN phases were accelerated by NVIDIA Quadro RTX 8000 GPU (48 GB memory).

3. Results

3.1. Initial Detection

Using the proposed method, we obtained the results for the initial detection before performing FP elimination. Figure 4 shows a lesion detected in the initial detection process (a−d) and an example of a missed lesion (e,f). The right image in Figure 4c,d shows those lesions that were missed by our previous technique but detected by the proposed method. As a result of the automated detection of all 1741 images using the cross-validation method, lesions were detected in 491 out of 533 images that contained lesions, while no lesions were detected in 42 images. When the detection sensitivity was evaluated on a lesion basis, the presence of GC was detected in at least one image in 98.9% (93/94) of patients, while 1.1% (1/94) of GCs were not detected in any patient. FPs were detected in 42 of 1208 images in the healthy group, resulting in an FP count of 0.035 per image.

3.2. False Positive Reduction

Box classification was performed on 533 images (491 true positives and 42 FPs) detected in the initial detection to eliminate the FPs. Figure 5 shows an example of a cropped image to be given to the CNN for FP reduction. Table 2 shows the detection sensitivity and the numbers of FPs per image and per lesion when an FP reduction was performed with six different CNN architectures. DenseNet169 showed the highest ability to eliminate FPs. Examples of FPs that could be removed by DenseNet169 and those that could not are shown in Figure 6. The results of the calculation of Di and Ji for GC cases are shown in Table 3.

Table 2. Comparison of CNN architectures for false positive reduction.

Classifier	Detection Sensitivity		False Positives/Image
	Lesion-Based	**Image-Based**	
None	0.989	0.942	0.0348
VGG16	0.989	0.942	0.0348
ResNet50	0.989	0.932	0.0281
DenseNet121	0.978	0.916	0.0273
DenseNet169	**0.989**	**0.897**	**0.0108**
DenseNet201	0.989	0.901	0.0240

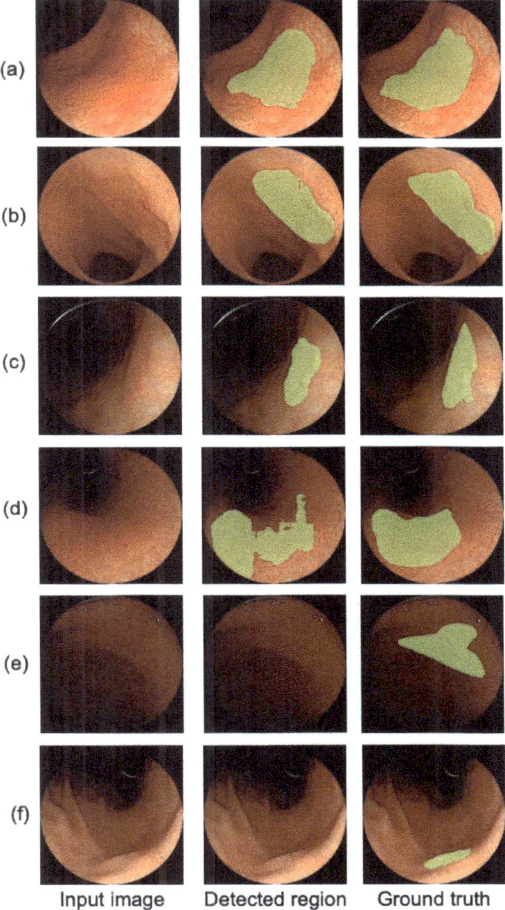

Figure 4. Lesions detected and missed in the initial detection process: (**a**,**b**) correctly detected by both the previous and the proposed method; (**c**,**d**) correctly detected only by the proposed method; (**e**,**f**) missed by the proposed method.

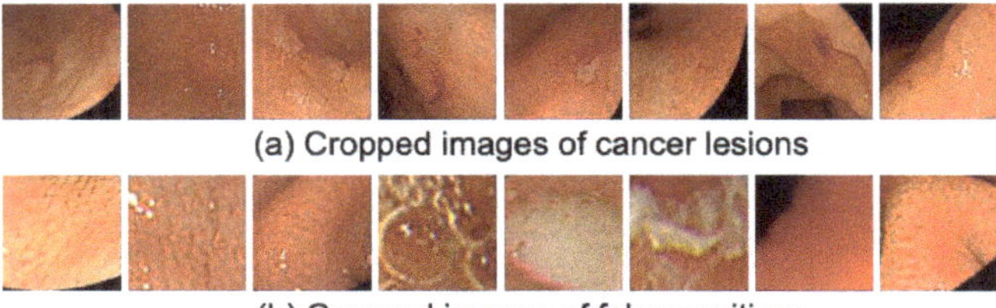

Figure 5. Example of the cropped images for false positive reduction: (**a**) Cropped images of cancer lesions; (**b**) Cropped images of false positives.

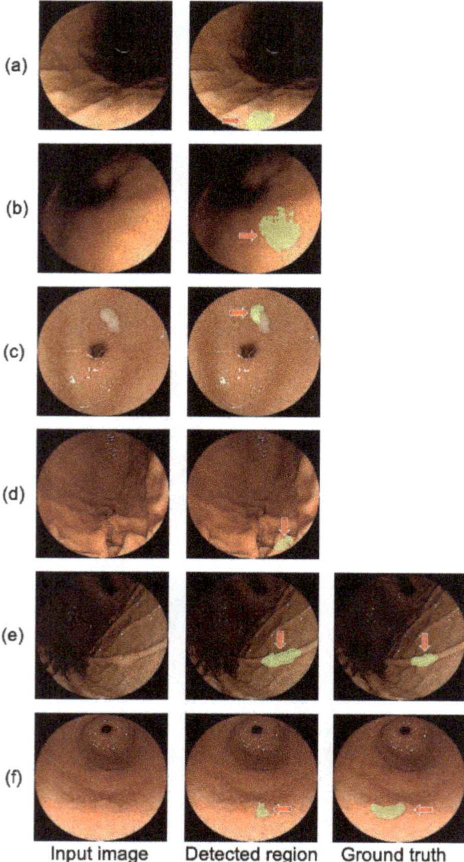

Figure 6. Examples of false positive reduction: (**a**,**b**) false positives correctly eliminated by the proposed method; (**c**,**d**) FPs that were not eliminated by the proposed method; (**e**,**f**) gastric cancer lesions eliminated by false positive reduction. Arrows indicate the labeled regions.

Table 3. Evaluation results of Di and Ji for cancer segmentation.

	Evaluation Using All Gastric Cancer Cases		Evaluation Using the Detected Gastric Cancer Cases	
	Di	Ji	Di	Ji
Previous method [16]	0.542	0.371	0.720	0.494
Proposed method	**0.555**	**0.427**	0.602	0.463

4. Discussion

In this study, we proposed a U-Net RCNN that combines U-Net and an FP reduction method for object detection and performs automatic detection of GC cases. Using the output images of U-Net, individual candidate regions were recognized by conventional thresholding and labeling techniques and bounding boxes were obtained. To eliminate FPs, candidate regions were classified as true GC cases and FPs by CNN.

The lesion-based sensitivity for initial detection by this method was 0.989, while the number of FPs per image was 0.035, which was much better than the previous study (sensitivity, 0.957; number of FPs per image, 0.105). The Mask R-CNN introduced in our previous method was able to accurately detect visually distinct objects such as unevenness due to the principles of the object detection model; however, it was difficult to detect subtle changes in the mucosal surface. On the other hand, U-Net, which was employed in this study, could analyze local regions in an image. A detailed analysis of the gastric mucosa in endoscopic images was performed and patterns that differed from normal were accurately recognized.

In the second stage, we compared the performance of six different CNN architectures and found that DenseNet169 showed the best performance, reducing FPs by approximately 30% to 0.011, while maintaining a lesion-based detection sensitivity of 0.989.

When evaluated in an image-based manner, the detection sensitivity dropped by approximately 4% from 0.942 to 0.897. As shown in Figure 5e,f, most of the images that remained undetected were taken from angles and distances that were difficult to see, and other images in the case were able to compensate for the detection.

The accuracy of extracting the invasive region of GC was evaluated by Di and Ji and the results were 0.55 and 0.42, respectively, for all GC images; and 0.60 and 0.46, respectively, when the evaluation was limited to the correctly detected images. The proposed method was more accurate than the previous study using Mask R-CNN when evaluating all GC images, while the previous study was more accurate when evaluating only the detected regions. This indicates that our method may detect subtle lesions but is not able to extract the exact shapes. To improve the extraction accuracy, it is necessary to improve the CNN model used for the initial detection and to add post-processing, such as region growing, to the extracted images.

Because the proposed method provides a sensitivity of 98% in detecting GC while keeping FPs at an acceptable level, it may be useful for maintaining high examination accuracy in screening for GC by covering differences in the experience of physicians.

Although we could not compare our results accurately because a different dataset was used, the proposed method using U-Net and FP reduction techniques had a better detection sensitivity than those in previous studies using a SSD [14] and Mask R-CNN [16]. Furthermore, a previous study using SSD detected lesions with a bounding box, whereas the proposed method segments the GC regions. The detection and segmentation capabilities of the proposed method are significantly improved compared to the previous methods.

The major limitation of the proposed method is the small number of images. Training and evaluation of the proposed method were carried out using the data collected at a single facility only for comparison with our previous method. We plan to expand the dataset by including data from external facilities.

5. Conclusions

In this study, we developed a deep learning model that can accurately detect the presence of GC and its invasive area using endoscopic images. In this paper, as a deep learning model, we proposed a novel U-Net R-CNN that combines the U-Net segmentation process with a CNN for image classification to eliminate FPs. As a result of the evaluation using the endoscopic images of early-stage GC and healthy subjects, the proposed method showed a higher detection ability than the previous techniques. These results indicate that our method is effective for the automated detection of early GC in endoscopy.

Author Contributions: Conceptualization, A.T., T.S., Y.H., K.S. and H.F.; data curation, T.S. and H.Y.; methodology, A.T.; software, A.T.; validation, T.S., A.T., and H.Y.; investigation, A.T., T.S., and H.Y.; writing—original draft preparation, A.T. and T.S.; writing—review and editing, Y.H., K.S. and H.F.; visualization, A.T.; project administration, A.T. and T.S. All authors have read and agreed to the published version of the manuscript.

Funding: This research received no external funding.

Institutional Review Board Statement: This study was approved by the Ethical Review Committee of Fujita Health University (HM17-226) and carried out in accordance with the World Medical Association's Declaration of Helsinki.

Informed Consent Statement: Patients' informed consent was obtained in the form of opt-out in the endoscopic center of Fujita Health University Hospital and all data were anonymized.

Data Availability Statement: The source code and additional information used to support the findings of this study will be available from the corresponding author upon request.

Conflicts of Interest: The authors declare no conflict of interest.

References

1. Fitzmaurice, C.; Akinyemiju, T.F.; Al Lami, F.H.; Alam, T.; Alizadeh-Navaei, R.; Allen, C.; Alsharif, U.; Alvis-Guzman, N.; Amini, E.; Anderson, B.O.; et al. Global, regional, and national cancer incidence, mortality, years of life lost, years lived with disability, and disability-adjusted life-years for 29 cancer groups, 1990 to 2016: A systematic analysis for the global burden of disease study global burden of disease cancer collaboration. *JAMA Oncol.* **2018**, *4*, 1553–1568. [PubMed]
2. Karger Publishers [Internet]. GLOBOCAN 2012: Estimated Cancer Incidence, Mortality, and Prevalence Worldwide in 2012. Available online: http://globocan.iarc.fr/Pages/fact_sheets_cancer.aspx (accessed on 31 October 2021).
3. Tashiro, A.; Sano, M.; Kinameri, K.; Fujita, K.; Takeuchi, Y. Comparing mass screening techniques for gastric cancer in Japan. *World J. Gastroenterol.* **2006**, *12*, 4873–4874. [PubMed]
4. Toyoizumi, H.; Kaise, M.; Arakawa, H.; Yonezawa, J.; Yoshida, Y.; Kato, M.; Yoshimura, N.; Goda, K.; Tajiri, H. Ultrathin endoscopy versus high-resolution endoscopy for diagnosing superficial gastric neoplasia. *Gastrointest. Endosc.* **2009**, *70*, 240–245. [CrossRef] [PubMed]
5. Teramoto, A.; Tsukamoto, T.; Yamada, A.; Kiriyama, Y.; Imaizumi, K.; Saito, K.; Fujita, H. Deep learning approach to classification of lung cytological images: Two-step training using actual and synthesized images by progressive growing of generative adversarial networks. *PLoS ONE* **2020**, *15*, e0229951. [CrossRef] [PubMed]
6. Yan, K.; Cai, J.; Zheng, Y.; Harrison, A.P.; Jin, D.; Tang, Y.B.; Tang, Y.X.; Huang, L.; Xiao, J.; Lu, L. Learning from Multiple Datasets with Heterogeneous and Partial Labels for Universal Lesion Detection in CT. *arXiv* **2020**, arXiv:2009.02577. [CrossRef] [PubMed]
7. Sahiner, B.; Pezeshk, A.; Hadjiiski, L.M.; Wang, X.; Drukker, K.; Cha, K.H.; Summers, R.M.; Giger, M.L. Deep learning in medical imaging and radiation therapy. *Med. Phys.* **2019**, *46*, e1–e36. [CrossRef] [PubMed]
8. Toda, R.; Teramoto, A.; Tsujimoto, M.; Toyama, H.; Imaizumi, K.; Saito, K.; Fujita, H. Synthetic CT Image Generation of Shape-Controlled Lung Cancer using Semi-Conditional InfoGAN and Its Applicability for Type Classification. *Int. J. Comput. Assist. Rad. Surg.* **2021**, *16*, 241–251. [CrossRef] [PubMed]
9. Tsujimoto, M.; Teramoto, A.; Dosho, M.; Tanahashi, S.; Fukushima, A.; Ota, S.; Inui, Y.; Matsukiyo, R.; Obama, Y.; Toyama, H. Automated classification of increased uptake regions in bone SPECT/CT images using three-dimensional deep convolutional neural network. *Nucl. Med. Commun.* **2021**, *42*, 877–883. [PubMed]
10. Teramoto, A.; Fujita, H.; Yamamuro, O.; Tamaki, T. Automated detection of pulmonary nodules in PET/CT images: Ensemble false-positive reduction using a convolutional neural network technique. *Med. Phys.* **2016**, *43*, 2821–2827. [CrossRef] [PubMed]
11. Shichijo, S.; Endo, Y.; Aoyama, K.; Takeuchi, Y.; Ozawa, T.; Takiyama, H.; Matsuo, K.; Fujishiro, M.; Ishihara, S.; Ishihara, R.; et al. Application of convolutional neural networks for evaluating Helicobacter pylori infection status on the basis of endoscopic images. *Scand. J. Gastroenterol.* **2019**, *54*, 158–163. [CrossRef] [PubMed]
12. Li, L.; Chen, Y.; Shen, Z.; Zhang, X.; Sang, J.; Ding, Y.; Yang, X.; Li, J.; Chen, M.; Jin, C.; et al. Convolutional neural network for the diagnosis of early gastric cancer based on magnifying narrow band imaging. *Gastric Cancer.* **2020**, *23*, 126–132. [CrossRef] [PubMed]
13. Zhang, X.; Hu, W.; Chen, F.; Liu, J.; Yang, Y.; Wang, L.; Duan, H.; Si, J. Gastric precancerous diseases classification using CNN with a concise model. *PLoS ONE* **2017**, *12*, e0185508. [CrossRef] [PubMed]
14. Hirasawa, T.; Aoyama, K.; Tanimoto, T.; Ishihara, S.; Shichijo, S.; Ozawa, T.; Ohnishi, T.; Fujishiro, M.; Matsuo, K.; Fujisaki, J.; et al. Application of artificial intelligence using a convolutional neural network for detecting gastric cancer in endoscopic images. *Gastric Cancer* **2018**, *21*, 653–660. [CrossRef] [PubMed]
15. Sakai, Y.; Takemoto, S.; Hori, K.; Nishimura, M.; Ikematsu, H.; Yano, T.; Yokota, H. Automatic detection of early gastric cancer in endoscopic images using a transferring convolutional neural network. In Proceedings of the 2018 40th Annual International Conference of the IEEE Engineering in Medicine and Biology Society (EMBC), Honolulu, HI, USA, 18–21 July 2018; pp. 4138–4141.
16. Shibata, T.; Teramoto, A.; Yamada, H.; Ohmiya, N.; Saito, K.; Fujita, H. Automated Detection and Segmentation of Early Gastric Cancer from Endoscopic Images Using Mask R-CNN. *Appl. Sci.* **2020**, *10*, 3842. [CrossRef]
17. Yagi, K.; Nakamura, A.; Sekine, A. Characteristic endoscopic and magnified endoscopic findings in the normal stomach without Helicobacter pylori infection. *J. Gastroenterol. Hepatol.* **2002**, *17*, 39–45. [CrossRef] [PubMed]
18. Ronneberger, O.; Fischer, P.; Brox, T. U-Net: Convolutional Networks for Biomedical Image Segmentation. *Lect. Notes. Comput. Sci.* **2015**, *9351*, 234–241.
19. Kingma, D.P.; Ba, J. Adam: A Method for Stochastic Optimization. *arXiv* **2017**, arXiv:1412.6980.
20. Otsu, N. A threshold selection method from gray-level histograms. *IEEE Trans. Syst. Man Cybern.* **1979**, *9*, 62–66. [CrossRef]
21. Simonyan, K.; Zisserman, A. Very deep convolutional networks for large-scale image recognition. *arXiv* **2015**, arXiv:1409.1556.
22. Szegedy, C.; Liu, W.; Jia, Y.; Sermanet, P.; Reed, S.; Anguelov, D.; Erhan, D.; Vanhoucke, V.; Rabinovich, A. Going deeper with convolutions. In Proceedings of the IEEE Conference on Computer Vision and Pattern Recognition (CVPR), Boston, MA, USA, 7–12 June 2015; pp. 1–9.

23. He, K.; Zhang, X.; Ren, S.; Sun, J. Deep residual learning for image recognition. In Proceedings of the IEEE Conference on Computer Vision and Pattern Recognition (CVPR), Las Vegas, NV, USA, 27–30 June 2016; pp. 770–778.
24. Huang, G.; Liu, Z.; van der Maaten, L.; Weinberger, K.Q. Densely connected convolutional networks. In Proceedings of the IEEE Conference on Computer Vision and Pattern Recognition (CVPR), Honolulu, HI, USA, 21–26 July 2017; pp. 4700–4708.
25. Efron, B. Estimating the error rate of a prediction rule: Improvement on cross-validation. *J. Am. Stat. Assoc.* **1983**, *78*, 316–331. [CrossRef]

Article

Backdoor Attacks to Deep Neural Network-Based System for COVID-19 Detection from Chest X-ray Images

Yuki Matsuo and Kazuhiro Takemoto *

Department of Bioscience and Bioinformatics, Kyushu Institute of Technology, Iizuka, Fukuoka 820-8502, Japan; matsuo.yuki678@mail.kyutech.jp
* Correspondence: takemoto@bio.kyutech.ac.jp; Tel.: +81-948-29-7822

Abstract: Open-source deep neural networks (DNNs) for medical imaging are significant in emergent situations, such as during the pandemic of the 2019 novel coronavirus disease (COVID-19), since they accelerate the development of high-performance DNN-based systems. However, adversarial attacks are not negligible during open-source development. Since DNNs are used as computer-aided systems for COVID-19 screening from radiography images, we investigated the vulnerability of the COVID-Net model, a representative open-source DNN for COVID-19 detection from chest X-ray images to backdoor attacks that modify DNN models and cause their misclassification when a specific trigger input is added. The results showed that backdoors for both non-targeted attacks, for which DNNs classify inputs into incorrect labels, and targeted attacks, for which DNNs classify inputs into a specific target class, could be established in the COVID-Net model using a small trigger and small fraction of training data. Moreover, the backdoors were effective for models fine-tuned from the backdoored COVID-Net models, although the performance of non-targeted attacks was limited. This indicated that backdoored models could be spread via fine-tuning (thereby becoming a significant security threat). The findings showed that emphasis is required on open-source development and practical applications of DNNs for COVID-19 detection.

Keywords: deep neural networks; medical imaging; backdoor attacks; security and privacy; COVID-19

Citation: Matsuo, Y.; Takemoto, K. Backdoor Attacks to Deep Neural Network-Based System for COVID-19 Detection from Chest X-ray Images. *Appl. Sci.* **2021**, *11*, 9556. https://doi.org/10.3390/app11209556

Academic Editor: Kyungtae Kang

Received: 14 September 2021
Accepted: 13 October 2021
Published: 14 October 2021

Publisher's Note: MDPI stays neutral with regard to jurisdictional claims in published maps and institutional affiliations.

1. Introduction

Deep neural networks (DNNs) demonstrate high performance in image recognition. Hence, they promise to achieve faster and more reliable decision-making in clinical environments as diagnostic medical imaging systems [1] since their diagnostic performance is high and equivalent to that of health care professionals [2]. For emerging infectious diseases such as the coronavirus disease 2019 (COVID-19) [3], DNNs are expected to effectively facilitate the screening of patients to reduce the spread of the epidemic. For instance, positive real-time polymerase chain reaction tests are generally used for COVID-19 screening [4]. However, they are often time-consuming and laborious and involve complicated manual processes. Thus, chest X-ray imaging has become an alternative screening method [5,6]. However, it is difficult to detect COVID-19 cases from chest X-ray images since visual differences in images between COVID-19 and non-COVID-19 pneumonias are subtle. Only a few expert radiologists have accurately detected COVID-19 from chest X-ray images, forming a bottleneck for faster screening based on radiographic images. DNNs can overcome this limitation due to the fact that they exhibit high performance for pneumonia classification based on chest X-ray images [7]. DNNs are now used to support radiologists in achieving a rapid and accurate interpretation of radiographic images for COVID-19 screening [8–15].

Specifically, the COVID-Net open-source initiative [8] demonstrates remarkable results. COVID-Net is a deep convolutional neural network designed to detect COVID-19 cases from chest X-ray images and is one of the first open-source network designs that detects COVID-19. To date, computer-based systems in medical science have generally been developed using closed sources in terms of security. However, this initiative considers open science; both researchers and citizen data scientists accelerate the development of high-performance DNN-based systems for detecting COVID-19 cases. Inspired by COVID-Net models, several researchers [16–18] have proposed DNN-based systems for COVID-19 screening from chest X-ray images. Moreover, large-scale datasets of chest radiography images of COVID-19 have been constructed [8,9,19,20]. Such open-source projects are encouraging not only for developing high-performance DNN solutions, but also for ensuring transparency and reproducibility in DNN models [21], although only deep learning models (model weights) may be provided [22] as an alternative to sharing patient data with regard to preserving patient privacy [23].

However, adversarial attacks hinder the development of open-source DNNs. In particular, DNNs are vulnerable to adversarial examples [24–26], which are input images contaminated with specific small perturbations that cause misclassifications by DNNs. Adversarial examples include evasion attacks in adversarial attacks. Many evasion attack methods (i.e., methods for generating adversarial examples) have been proposed, such as the fast gradient sign method [24] and DeepFool [27]). Since disease diagnosis involves high-stake decisions, adversarial attacks can cause serious security problems [28] and various social problems [29]. Thus, the vulnerability of DNNs to evasion attacks has been investigated in medical imaging [29,30]. For COVID-19 detection, adversarial attacks may hinder strategies for public health (i.e., minimizing the spread of the pandemic) and the economy. For open-source DNNs such as the COVID-Net model, adversaries can easily generate adversarial examples since they can access the model parameters (the model weights and gradient of the loss function) and training images. We previously [31] demonstrated that universal adversarial perturbation (UAP) [32,33], an evasion attack using a single (input image agnostic) perturbation can fail most classification tasks of the COVID-Net model.

Nevertheless, backdoor attacks [34], which are different types of adversarial attacks, must be considered to obtain a more comprehensive understanding of security threats to open-source DNNs since previous studies have only focused on evasion attacks (i.e., manipulating inputs to cause DNN misclassifications). In backdoor attacks, a backdoor is established in DNN models (i.e., model poisoning) to misclassify them; specifically, backdoor attacks are performed by fine-tuning existing DNN models with contaminated data that are generated by assigning backdoor triggers (e.g., a pixel pattern that appears in the corner of the images) and incorrect labels to a small fraction of the original data. In this case, backdoored DNN models correctly classify inputs without triggers into their actual labels. However, they incorrectly predict the actual labels for inputs with triggers. Depending on the manner in which incorrect labels are assigned to contaminated data, both non-targeted attacks, for which DNNs classify inputs into incorrect labels, and targeted attacks, for which DNNs classify inputs into a specific target class, can be implemented. It is difficult to immediately discriminate whether backdoors are established in DNN models since DNN models appear to function correctly for inputs without backdoor triggers and exhibit complex architectures. Open-source software development relies on collaboration among researchers, engineers, citizen data scientists, etc. and it may be outsourced. In this situation, an unspecified number of people can be involved in development. Thus, anyone can establish a backdoor in DNN models via the above procedures. Moreover, it is difficult to determine who establishes the backdoor. Backdoor attacks are a serious security threat for open-source software development [34]. Therefore, they have been evaluated in handwritten digit recognition tasks, traffic sign detection tasks, and well-used sources for pretrained DNN models [34]. However, the vulnerability of existing open-source software in medical imaging (e.g., the COVID-Net model) to backdoor attacks has not been evaluated comprehensively at present, although a previous study [35] considered backdoor attacks on medical imaging based on DNN models trained by the authors themselves.

This study's aim is to evaluate the vulnerability of the COVID-Net model, a representative open-source software used in medical imaging, for backdoor attacks. Specifically, we evaluate whether backdoors for non-targeted and targeted attacks can be established in the

COVID-Net models. Moreover, the effectiveness of the backdoors in DNN models fine-tuned from backdoored models is analyzed. Backdoor attacks cause a significant problem when fine-tuned models are obtained from backdoored models. In medical imaging, users often consider obtaining highly accurate DNN models by fine-tuning pretrained models with their own datasets since the amount of medical image data is often limited [1]. Users may perceive that they have obtained highly accurate fine-tuned DNN models from backdoored models since the models function correctly for clean inputs. However, adversaries can foil or control the tasks of fine-tuned DNN models using backdoor triggers. Therefore, we evaluated whether the backdoor triggers enabled non-targeted and targeted attacks for DNN models fine-tuned from backdoored models.

2. Materials and Methods

2.1. COVID-Net Model and Chest X-ray Images

We obtained a COVID-Net model and chest X-ray images based on a previous study [31]. In particular, the COVIDNet-CXR4-A model was downloaded from the GitHub repository on the COVID-Net Open Source Initiative (https://github.com/lindawangg/COVID-Net) on 20 November 2020. This model was selected since its prediction accuracy was the highest (94.3%) at that time. Moreover, we downloaded the COVIDx5 dataset, which was constructed using several open-source chest radiography datasets, on 19 November 2020, following the description in the COVID-Net repository (see https://github.com/lindawangg/COVID-Net/blob/master/docs/COVIDx.md (accessed on 19 November 2020) for details). In particular, the dataset consisted of COVID-19 image data collection [36], COVID-19 Radiography Database [37,38], hospital-scale chest X-ray database (ChestX-Ray8) [39], The Radiological Society of North America International COVID-19 Open Radiology Database (RICORD) [40], etc. The images were in grayscale with a pixel resolution of 480×480 pixels and a pixel intensity ranging between 0 pixels and 255 pixels. The chest X-ray images in the dataset were classified into three classes: normal (no pneumonia), pneumonia (non-COVID-19 pneumonia; e.g., viral and bacterial pneumonia), and COVID-19 (COVID-19 viral pneumonia). The COVIDx5 dataset comprised 13,958 training images (7966 normal, 5475 pneumonia, and 517 COVID-19) and 300 test images (100 images per class).

The COVIDx5 dataset was classified into two datasets: Datasets 1 and 2. Dataset 1 contained 6978 training images (3983 normal, 2737 pneumonia, and 258 COVID-19) and 150 test images (50 images per class), which were randomly selected from the COVIDx5 dataset. These training and test images were used to establish a backdoor in the COVID-Net model (i.e., to generate a backdoor COVID-Net model) and to evaluate the performance of the backdoor attacks. The remainder of the COVIDx5 dataset corresponded to Dataset 2, which contained 6980 training images (3983 normal, 2738 pneumonia, and 259 COVID-19) and 150 test images (50 images per class). These training and test images were used to obtain a fine-tuned model from the backdoor COVID-Net model and to evaluate the performance of backdoor attacks on the fine-tuned model.

2.2. Backdoor Attacks

The procedure for establishing a backdoor in the COVID-Net model was based on a previous study [34]. To obtain a contaminated training dataset, a backdoor trigger was applied to 698 (~10%) images (398 normal, 273 pneumonia, and 25 COVID-19) that were randomly selected from the training images in Dataset 1. The trigger was set to a square measuring 5×5 pixels (~1% height and width of the images) and a pixel intensity of 250, and it was placed at the lower right corner [near pixel coordinated (398, 398)] of the images. For each image x, image x_t (the trigger) was generated by applying the trigger to x using the matrix of a 480×480 image mask, m, which assumed a value of 1 at the coordination where the trigger was located, and 0 otherwise: $x_t = \tau(x) = x \circ (1 - m) + 250m$, where $\circ$ indicated the element-wise product and **1** was the 480×480 matrix in which all elements were 1. Figure 1 shows the examples of normal, pneumonia, and COVID-19 images, with and without the trigger. Furthermore, incorrect labels were assigned to the images with the trigger. For non-targeted attacks, we assigned pneumonia, COVID-19, and normal labels to normal images, pneumonia images, and COVID-19 images, respectively. For targeted attacks, a target label was assigned to all the images.

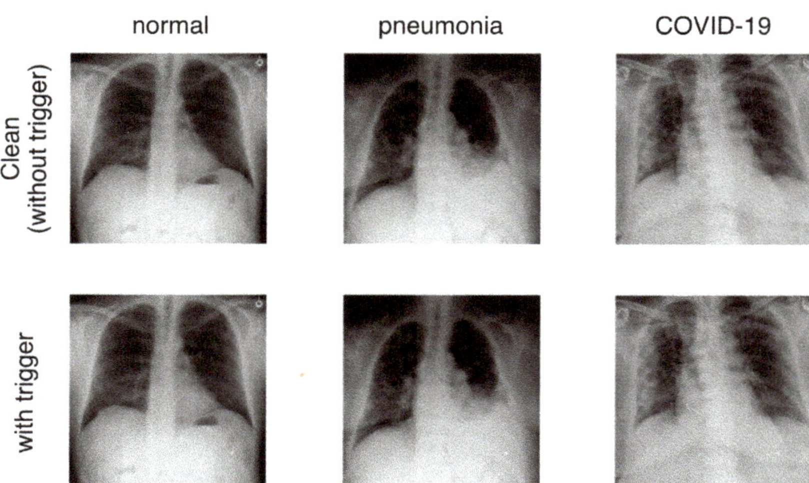

Figure 1. Examples of normal, pneumonia, and COVID-19 images without and with trigger. Example images were randomly selected per class.

Using the contaminated training dataset, we fine-tuned the COVID-Net model with batch sizes of 32 and 50 epochs. The other settings (e.g., learning rate and optimizer) were the same as those used for training the original COVID-Net model.

2.3. Model Fine-Tuned from Backdoor Model

We obtained a fine-tuned model for COVID-19 detection using the backdoor COVID-Net model. Specifically, using the training images in Dataset 2, we fine-tuned the backdoor model with batch sizes of 32 and 20 epochs. The other settings (e.g., learning rate and optimizer) were the same as those used for training the original COVID-Net model.

2.4. Evaluating Performance of Backdoor Attacks

The performance of the backdoor attacks with the trigger was evaluated based on the attack success rates. Specifically, based on previous studies [31,41], we used the fooling rate R_f and targeted attack success rate R_s to evaluate the performance of non-targeted and targeted attacks, respectively. Let $C(x)$ and y_x be an output (class or label) of a classifier (DNN) and the actual label for an input image x, respectively; R_f was defined as the fraction of cases in which the labels predicted from images with the trigger differed from those from their images without the trigger for all images in set X: $R_f = |X|^{-1} \sum_{x \in X} \mathbb{I}(y_x \neq C(\tau(x)))$, where $\mathbb{I}(A)$ was 1 if condition A was true, and 0 otherwise. R_s was defined as the ratio of images with the trigger classified into a target class K to all images in set X: $R_s = |X|^{-1} \sum_{x \in X} \mathbb{I}(C(\tau(x)) = K)$. To evaluate the change in the predicted labels for each class due to the trigger, confusion matrices were obtained. R_f, R_s, and the confusion matrices were computed using the test images in Datasets 1 and 2 to evaluate the performance of the backdoor attacks on the backdoor model and the model fine-tuned from the backdoor model, respectively.

3. Results

First, we investigated whether backdoors for non-targeted and targeted attacks could be established in the COVID-Net model. The prediction accuracies (Table 1) and confusion matrices (the upper panels in Figure 2) indicated that the backdoor models of the COVID-Net model demonstrated high prediction performance for clean images (i.e., images without the trigger (see the upper panels in Figure 1)), although their accuracies were slightly lower than those of the original COVID-Net model (e.g., the backdoor models for targeted attacks tended to classify some of the clean COVID-19 images as pneumonia (see the upper panels in Figure 2a–c)). However, the backdoor models classified the images with the trigger into target labels for targeted attacks and incorrect labels for non-targeted attacks (see bottom panels in Figure 2). The attack success rates (R_s or R_f) were between 85% and 100% (Table 1). The results indicated that backdoors were established in the COVID-Net model using a small trigger.

Table 1. Attack success rates (R_s for targeted attacks and R_f for non-targeted attacks; %) for backdoored COVID-Net models and prediction accuracies (%) of backdoored models on clean images.

Attack Type		Attack Success Rate (R_s or R_f)	Accuracy
Targeted	normal	99.3%	88.7%
	pneumonia	99.3%	78.7%
	COVID-19	100.0%	87.3%
Non-targeted		86.7%	91.3%

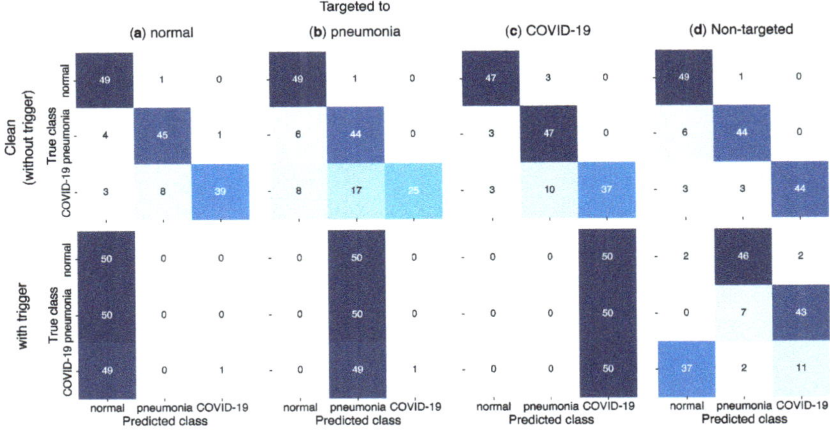

Figure 2. Confusion matrices for backdoored COVID-Net models on test images without any trigger (clean images; upper panels) and with trigger (bottom panels). Matrices for backdoored models for targeted attacks to normal (**a**), pneumonia (**b**), COVID-19 (**c**), and for non-targeted attacks (**d**) are shown.

Further, we evaluated whether backdoor attacks were effective for models fine-tuned from backdoored models. It was assumed that other users, not adversaries, obtained the fine-tuned models from the backdoored models using clean images, and used a publicly available DNN model to obtain their own models without knowing whether a backdoor was established in the DNN model. The prediction accuracies (Table 2) and confusion matrices (the upper panels in Figure 3) indicated that the fine-tuned models demonstrated high prediction performance for the clean images, and that their prediction accuracies were almost similar to those of the original COVID-Net model. Nevertheless, the backdoor attacks were effective in the fine-tuned models. Specifically, the success rates (R_s) for targeted attacks were between 60% and 80% (Table 2). However, the R_s of the fine-tuned models were lower than those of the backdoored models. In particular, the normal and COVID-19 images were difficult to misclassify, although the trigger was added to the images (the bottom panels in Figure 3a–c). Moreover, the performance of the non-targeted attacks was limited. In particular, R_f was approximately 10% (see the bottom panel in Figure 3d).

Table 2. Attack success rates (R_s for targeted attacks and R_f for non-targeted attacks; %) for fine-tuned models from backdoored COVID-Net models and prediction accuracies (%) of fine-tuned models on clean images.

	Attack Type		Attack Success Rate (R_s or R_f)	Accuracy
		normal	80.7%	91.3%
Targeted		pneumonia	60.0%	96.0%
		COVID-19	73.3%	90.7%
	Non-targeted		86.7%	11.3%

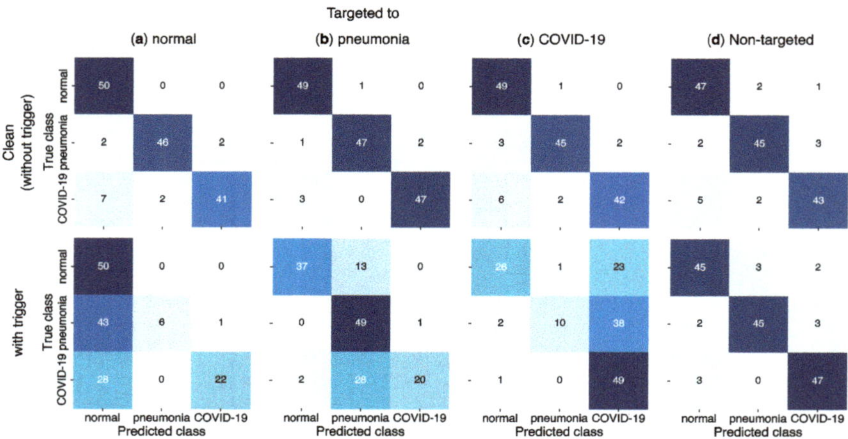

Figure 3. Confusion matrices for models fine-tuned from backdoored COVID-Net models on test images without any trigger (clean images; upper panels) and with trigger (bottom panels). Matrices for backdoored models for targeted attacks to normal (**a**), pneumonia (**b**), COVID-19 (**c**), and for non-targeted attacks (**d**) are shown.

4. Discussion

The results (Table 1 and Figure 2) show that the backdoors for both the non-targeted and targeted attacks were easily established in the COVID-Net model by assigning a small trigger and incorrect labels to a small fraction of training data. Similar to evasion attacks using UAPs [31], backdoor attacks also achieved high attack success rates (85% to 100%), indicating that the COVID-Net model was vulnerable to model poisoning. Users (e.g., developers except for adversaries) might not be easily detected, whereas the training data were contaminated due to the small number of training images with the trigger and incorrect labels. Hence, they might render the backdoor models publicly available. Other users fine-tuned the backdoored models using their training data to obtain their own DNN models for COVID-19 detection. Subsequently, fine-tuned models with high prediction performances were obtained (Table 2). Nonetheless, the backdoors for the targeted attacks remained effective for the fine-tuned models (Table 2 and Figure 3). The fine-tuned models would be used in real-world environments since they functioned correctly for inputs without a trigger. The spread of backdoor models via fine-tuning might pose a significant security threat. In particular, adversaries could easily attack several fine-tuned models from the backdoored models using typical triggers to cause both false positives and negatives in COVID-19 diagnosis. This might cause problems for public health and the economy, as mentioned in a previous study [31]. False positives in the diagnosis due to backdoor attacks might cause undesired mental stress in patients. False negatives in the diagnosis due to the attacks might have facilitated the spread of the pandemic. Moreover, backdoor attacks could be used to adjust the number of COVID-19 cases. Therefore, they might complicate the estimation of the number of COVID-19 cases. These disturbances due to backdoor attacks affected the individual and social awareness of COVID-19 (e.g., voluntary restraint and social distancing) and therefore hindered the spread of the pandemic.

However, backdoor attacks on the COVID-Net model were less effective. For the backdoor models, their prediction accuracies on clean images were slightly lower than those of the original COVID-Net model. In particular, some of the clean COVID-19 images were classified as pneumonia (Figure 1). This might be due to the fact that the visual differences in chest X-ray images between COVID-19 and non-COVID-19 pneumonia were insignificant. The decision boundary between COVID-19 and pneumonia might have been altered due to the backdoor trigger. For the fine-tuned models, the performance of backdoor attacks was lower than that of the backdoored models. Specifically, normal and COVID-19 images with the trigger were difficult to misclassify (Figure 2a–c). This might be due to the significant visual differences in chest X-ray images between non-pneumonia and COVID-19 pneumonia. The decision boundary between normal and COVID-19 that was altered due to the backdoor trigger might have returned to the original state since fine-tuning was performed using clean images. Furthermore, the backdoor for non-targeted attacks was not effective for the fine-tuned model. This might be due to the fact that it was difficult to assign incorrect labels to the images with the trigger. In particular, the decision boundary for each class was altered using backdoor triggers. However, this alteration might have been difficult when using only a single trigger.

Explainability might be a useful indicator for determining whether backdoors were established in DNN models. Gradient class activation mapping (Grad-CAM) [42] was useful in this context [43]. It provided saliency maps that indicated the importance of each pixel in the input images for the model outputs (i.e., prediction results) using the gradients of the outputs with respect to activation functions until the final convolution layer. The saliency maps of the backdoored models differed from those of the clean models. Specifically, pixels at unexpected coordinates (e.g., near a backdoor trigger) contributed to model predictions. Nwadike et al. [35] detected backdoor attacks on medical imaging using DNN models trained by themselves using Grad-CAM saliency maps, inspired by the fact that explainability techniques were typically used in medical imaging applications [44]. However, adversarial defenses against backdoor attacks based on explainability might be limited since explainability could be easily deceived [45]. Specifically, adversaries could fine-tune DNN models to allow explainability methods (e.g., Grad-CAM) to yield their desired saliency maps. Moreover, explainabiltiy-based defenses had failed to combat imperceptible backdoor attacks based on image warping [46] and physical reflection [47]. Adversarial attacks and defenses were cat-and-mouse games [29]. Hence, it might be difficult to defend against backdoor attacks.

The vulnerability to backdoor attacks demonstrated here was limited to the COVID-Net model. This was due to the fact that the number of reproducible open-source projects on DNN-based COVID-19 detection was limited at that time. However, we believed that vulnerability was a general property of DNN models, given that backdoor attacks were effective in DNN models for various types of classification tasks [34,35]. The vulnerability of other DNN models for COVID-19 detection to backdoor attacks needs to be further investigated; however, the procedures used here might be useful as a standard framework for evaluating the vulnerability of DNN models.

5. Conclusions

The vulnerability of the COVID-Net model, an open-source DNN, for backdoor attacks was demonstrated. Collaboration among researchers, engineers, and citizen data scientists were expected in open-source projects to accelerate the development of high-performance DNNs. However, the risk of backdoor attacks was inevitable. Although many DNN-based systems for COVID-19 detection were developed, the abovementioned risks were disregarded. Our findings highlighted that careful consideration is required in open-source development and practical applications of DNNs for COVID-19 detection.

Author Contributions: Conceptualization, K.T.; methodology, Y.M. and K.T.; software, Y.M.; validation, Y.M. and K.T.; formal analysis, Y.M. and K.T.; investigation, Y.M. and K.T.; resources, Y.M.; data curation, Y.M.; writing—original draft preparation, K.T.; writing—review and editing, Y.M. and K.T.; visualization, Y.M. and K.T.; supervision, K.T.; project administration, K.T.; funding acquisition, K.T. All authors have read and agreed to the published version of the manuscript.

Funding: This study was funded by JSPS KAKENHI (grant number JP21H03545).

Institutional Review Board Statement: Not applicable.

Informed Consent Statement: Not applicable.

Data Availability Statement: The code used here is available from the GitHub repository https://github.com/YukiM00/Backdoored-COVID-Net (accessed on 19 November 2020). The chest X-ray images used here are publicly available online (see https://github.com/lindawangg/COVID-Net/blob/master/docs/COVIDx.md (accessed on 19 November 2020) for details).

Conflicts of Interest: The authors declare no conflict of interest.

References

1. Litjens, G.; Kooi, T.; Bejnordi, B.E.; Setio, A.A.A.; Ciompi, F.; Ghafoorian, M.; van der Laak, J.A.W.M.; van Ginneken, B.; Sánchez, C.I. A survey on deep learning in medical image analysis. *Med. Image Anal.* **2017**, *42*, 60–88. [CrossRef]
2. Liu, X.; Faes, L.; Kale, A.U.; Wagner, S.K.; Fu, D.J.; Bruynseels, A.; Mahendiran, T.; Moraes, G.; Shamdas, M.; Kern, C.; et al. A comparison of deep learning performance against health-care professionals in detecting diseases from medical imaging: A systematic review and meta-analysis. *Lancet Digit. Health* **2019**, *1*, e271–e297. [CrossRef]
3. Dong, E.; Du, H.; Gardner, L. An interactive web-based dashboard to track COVID-19 in real time. *Lancet Infect. Dis.* **2020**, *20*, 533–534. [CrossRef]
4. Wang, D.; Hu, B.; Hu, C.; Zhu, F.; Liu, X.; Zhang, J.; Wang, B.; Xiang, H.; Cheng, Z.; Xiong, Y.; et al. Clinical characteristics of 138 hospitalized patients with 2019 novel coronavirus–infected pneumonia in Wuhan, China. *JAMA* **2020**, *323*, 1061. [CrossRef] [PubMed]
5. Ng, M.-Y.; Lee, E.Y.; Yang, J.; Yang, F.; Li, X.; Wang, H.; Lui, M.M.; Lo, C.S.-Y.; Leung, B.; Khong, P.-L.; et al. Imaging profile of the COVID-19 infection: Radiologic findings and literature review. *Radiol. Cardiothorac. Imaging* **2020**, *2*, e200034. [CrossRef] [PubMed]
6. Fang, Y.; Zhang, H.; Xie, J.; Lin, M.; Ying, L.; Pang, P.; Ji, W. Sensitivity of chest CT for COVID-19: Comparison to RT-PCR. *Radiology* **2020**, *296*, E115–E117. [CrossRef] [PubMed]
7. Kermany, D.S.; Goldbaum, M.; Cai, W.; Valentim, C.C.S.; Liang, H.; Baxter, S.L.; McKeown, A.; Yang, G.; Wu, X.; Yan, F.; et al. Identifying medical diagnoses and treatable diseases by image-based deep learning. *Cell* **2018**, *172*, 1122–1131.e9. [CrossRef] [PubMed]
8. Wang, L.; Lin, Z.Q.; Wong, A. COVID-Net: A tailored deep convolutional neural network design for detection of COVID-19 cases from chest X-ray images. *Sci. Rep.* **2020**, *10*, 19549. [CrossRef]
9. Zhang, K.; Liu, X.; Shen, J.; Li, Z.; Sang, Y.; Wu, X.; Cha, Y.; Liang, W.; Wang, C.; Wang, K.; et al. Clinically applicable AI system for accurate diagnosis, quantitative measurements and prognosis of COVID-19 pneumonia using computed tomography. *Cell* **2020**, *181*, 1423–1433.e11. [CrossRef]
10. Liu, S.; Shih, F.Y.; Zhong, X. Classification of chest X-ray images using novel adaptive morphological neural networks. *Int. J. Pattern Recognit. Artif. Intell.* **2021**, *35*, 2157006. [CrossRef]
11. Santosh, K.; Ghosh, S. Covid-19 imaging tools: How big data is big? *J. Med. Syst.* **2021**, *45*, 71. [CrossRef] [PubMed]
12. Das, D.; Santosh, K.C.; Pal, U. Truncated inception net: COVID-19 outbreak screening using chest X-rays. *Phys. Eng. Sci. Med.* **2020**, *43*, 915–925. [CrossRef] [PubMed]
13. Sadre, R.; Sundaram, B.; Majumdar, S.; Ushizima, D. Validating deep learning inference during chest X-ray classification for COVID-19 screening. *Sci. Rep.* **2021**, *11*, 16075. [CrossRef] [PubMed]
14. Mukherjee, H.; Ghosh, S.; Dhar, A.; Obaidullah, S.M.; Santosh, K.C.; Roy, K. Deep neural network to detect COVID-19: One architecture for both CT scans and chest X-rays. *Appl. Intell.* **2021**, *51*, 2777–2789. [CrossRef]
15. Stubblefield, J.; Hervert, M.; Causey, J.L.; Qualls, J.A.; Dong, W.; Cai, L.; Fowler, J.; Bellis, E.; Walker, K.; Moore, J.H.; et al. Transfer learning with chest X-rays for ER patient classification. *Sci. Rep.* **2020**, *10*, 20900. [CrossRef]

16. Farooq, M.; Hafeez, A. COVID-ResNet: A deep learning framework for screening of COVID19 from radiographs. *arXiv* **2020**, arXiv:2003.14395.
17. Afshar, P.; Heidarian, S.; Naderkhani, F.; Oikonomou, A.; Plataniotis, K.N.; Mohammadi, A. COVID-CAPS: A capsule network-based framework for identification of COVID-19 cases from X-ray Images. *Pattern Recognit. Lett.* **2020**, *138*, 638–643. [CrossRef]
18. Rahimzadeh, M.; Attar, A. A new modified deep convolutional neural network for detecting COVID-19 from X-ray images. *arXiv* **2020**, arXiv:2004.08052.
19. Zhao, J.; Zhang, Y.; He, X.; Xie, P. COVID-CT-Dataset: A CT scan dataset about COVID-19. *arXiv* **2020**, arXiv:2003.13865.
20. Cohen, J.P.; Morrison, P.; Dao, L. COVID-19 image data collection. *arXiv* **2020**, arXiv:2003.11597.
21. Haibe-Kains, B.; Adam, G.A.; Hosny, A.; Khodakarami, F.; Waldron, L.; Wang, B.; McIntosh, C.; Goldenberg, A.; Kundaje, A.; Greene, C.S.; et al. Transparency and reproducibility in artificial intelligence. *Nature* **2020**, *586*, E14–E16. [CrossRef] [PubMed]
22. Chang, K.; Balachandar, N.; Lam, C.; Yi, D.; Brown, J.; Beers, A.; Rosen, B.; Rubin, D.L.; Kalpathy-Cramer, J. Distributed deep learning networks among institutions for medical imaging. *J. Am. Med. Inform. Assoc.* **2018**, *25*, 945–954. [CrossRef] [PubMed]
23. Price, W.N.; Cohen, I.G. Privacy in the age of medical big data. *Nat. Med.* **2019**, *25*, 37–43. [CrossRef] [PubMed]
24. Goodfellow, I.J.; Shlens, J.; Szegedy, C. Explaining and harnessing adversarial examples. *arXiv* **2015**, arXiv:1412.6572.
25. Yuan, X.; He, P.; Zhu, Q.; Li, X. Adversarial examples: Attacks and defenses for deep learning. *IEEE Trans. Neural Netw. Learn. Syst.* **2019**, *30*, 2805–2824. [CrossRef] [PubMed]
26. Ortiz-Jimenez, G.; Modas, A.; Moosavi-Dezfooli, S.-M.; Frossard, P. Optimism in the face of adversity: Understanding and improving deep learning through adversarial robustness. *arXiv* **2020**, arXiv:2010.09624.
27. Moosavi-Dezfooli, S.-M.; Fawzi, A.; Frossard, P. DeepFool: A simple and accurate method to fool deep neural networks. In Proceedings of the 2016 IEEE Conference on Computer Vision and Pattern Recognition (CVPR), Las Vegas, NV, USA, 27–30 June 2016; pp. 2574–2582.
28. Kaissis, G.A.; Makowski, M.R.; Rückert, D.; Braren, R.F. Secure, privacy-preserving and federated machine learning in medical imaging. *Nat. Mach. Intell.* **2020**, *2*, 305–311. [CrossRef]
29. Finlayson, S.G.; Bowers, J.D.; Ito, J.; Zittrain, J.L.; Beam, A.L.; Kohane, I.S. Adversarial attacks on medical machine learning. *Science* **2019**, *363*, 1287–1289. [CrossRef]
30. Asgari Taghanaki, S.; Das, A.; Hamarneh, G. Vulnerability analysis of chest X-ray image classification against adversarial attacks. In *Understanding and Interpreting Machine Learning in Medical Image Computing Applications*; Springer: Cham, Switzerland, 2018; Volume 11038 LNCS, pp. 87–94. ISBN 9783030026271.
31. Hirano, H.; Koga, K.; Takemoto, K. Vulnerability of deep neural networks for detecting COVID-19 cases from chest X-ray images to universal adversarial attacks. *PLoS ONE* **2020**, *15*, e0243963. [CrossRef]
32. Moosavi-Dezfooli, S.M.; Fawzi, A.; Fawzi, O.; Frossard, P. Universal adversarial perturbations. In Proceedings of the 30th IEEE Conference on Computer Vision and Pattern Recognition (CVPR 2017), Honolulu, HI, USA, 21–26 July 2017; pp. 86–94. [CrossRef]
33. Hirano, H.; Takemoto, K. Simple iterative method for generating targeted universal adversarial perturbations. *Algorithms* **2020**, *13*, 268. [CrossRef]
34. Gu, T.; Liu, K.; Dolan-Gavitt, B.; Garg, S. BadNets: Evaluating backdooring attacks on deep neural networks. *IEEE Access* **2019**, *7*, 47230–47244. [CrossRef]
35. Nwadike, M.; Miyawaki, T.; Sarkar, E.; Maniatakos, M.; Shamout, F. Explainability matters: Backdoor attacks on medical imaging. In Proceedings of the AAAI 2021 Workshop: Trustworthy AI for Healthcare, Online, 9 February 2021.
36. Cohen, J.P.; Morrison, P.; Dao, L.; Roth, K.; Duong, T.Q.; Ghassemi, M. COVID-19 image data collection: Prospective predictions are the future. *arXiv* **2020**, arXiv:2006.11988.
37. Chowdhury, M.E.H.; Rahman, T.; Khandakar, A.; Mazhar, R.; Kadir, M.A.; Bin Mahbub, Z.; Islam, K.R.; Khan, M.S.; Iqbal, A.; Al Emadi, N.; et al. Can AI help in screening viral and COVID-19 pneumonia? *IEEE Access* **2020**, *8*, 132665–132676. [CrossRef]
38. Rahman, T.; Khandakar, A.; Qiblawey, Y.; Tahir, A.; Kiranyaz, S.; Bin Abul Kashem, S.; Islam, M.T.; Al Maadeed, S.; Zughaier, S.M.; Khan, M.S.; et al. Exploring the effect of image enhancement techniques on COVID-19 detection using chest X-ray images. *Comput. Biol. Med.* **2021**, *132*, 104319. [CrossRef] [PubMed]
39. Wang, X.; Peng, Y.; Lu, L.; Lu, Z.; Bagheri, M.; Summers, R.M. ChestX-Ray8: Hospital-scale chest X-ray database and benchmarks on weakly-supervised classification and localization of common thorax diseases. In Proceedings of the 30th IEEE Conference on Computer Vision and Pattern Recognition (CVPR 2017), Honolulu, HI, USA, 21–26 July 2017; pp. 3462–3471.
40. Tsai, E.B.; Simpson, S.; Lungren, M.P.; Hershman, M.; Roshkovan, L.; Colak, E.; Erickson, B.J.; Shih, G.; Stein, A.; Kalpathy-Cramer, J.; et al. The RSNA International COVID-19 Open Radiology Database (RICORD). *Radiology* **2021**, *299*, E204–E213. [CrossRef] [PubMed]
41. Hirano, H.; Minagi, A.; Takemoto, K. Universal adversarial attacks on deep neural networks for medical image classification. *BMC Med. Imaging* **2021**, *21*, 9. [CrossRef] [PubMed]
42. Selvaraju, R.R.; Cogswell, M.; Das, A.; Vedantam, R.; Parikh, D.; Batra, D. Grad-CAM: Visual explanations from deep networks via gradient-based localization. *Int. J. Comput. Vis.* **2020**, *128*, 336–359. [CrossRef]
43. Xu, K.; Liu, S.; Chen, P.-Y.; Zhao, P.; Lin, X. Defending against backdoor attack on deep neural networks. *arXiv* **2020**, arXiv:2002.12162.
44. Holzinger, A.; Langs, G.; Denk, H.; Zatloukal, K.; Müller, H. Causability and explainability of artificial intelligence in medicine. *WIREs Data Min. Knowl. Discov.* **2019**, *9*, e1312. [CrossRef] [PubMed]

45. Subramanya, A.; Pillai, V.; Pirsiavash, H. Fooling network interpretation in image classification. In Proceedings of the IEEE/CVF International Conference on Computer Vision (ICCV), Seoul, Korea, 27 October–2 November 2019.
46. Nguyen, T.A.; Tran, A.T. WaNet—Imperceptible warping-based backdoor attack. In Proceedings of the International Conference on Learning Representations, Virtual Event, 3–7 May 2021.
47. Liu, Y.; Ma, X.; Bailey, J.; Lu, F. Reflection backdoor: A natural backdoor attack on deep neural networks. In Proceedings of the European Conference on Computer Vision, Glasgow, UK, 23–28 August 2020; pp. 182–199.

Article

Instance-Based Learning Following Physician Reasoning for Assistance during Medical Consultation

Matías Galnares *, Sergio Nesmachnow * and Franco Simini *

Universidad de la República, Montevideo 11200, Uruguay
* Correspondence: matias.galnares@fing.edu.uy (M.G.); sergion@fing.edu.uy (S.N.); simini@fing.edu.uy (F.S.)

Abstract: This article presents an automatic system for modeling clinical knowledge to follow a physician's reasoning in medical consultation. Instance-based learning is applied to provide suggestions when recording electronic medical records. The system was validated on a real case study involving advanced medical students. The proposed system is accurate and efficient: $2.5\times$ more efficient than a baseline empirical tool for suggestions and two orders of magnitude faster than a Bayesian learning method, when processing a testbed of 250 clinical case types. The research provides a framework to implement a real-time system to assist physicians during medical consultations.

Keywords: computational intelligence; medical assistance; instance-based learning; healthcare; clinical decision support systems

Citation: Galnares, M.; Nesmachnow, S.; Simini, F. Instance-Based Learning Following Physician Reasoning for Assistance during Medical Consultation. *Appl. Sci.* **2021**, *13*, 5886. https://doi.org/10.3390/app11135886

Academic Editors: Kyungtae Kang, Junggab Son and Hyo-Joong Suh

Received: 29 May 2021
Accepted: 21 June 2021
Published: 24 June 2021

Publisher's Note: MDPI stays neutral with regard to jurisdictional claims in published maps and institutional affiliations.

1. Introduction

The search for better medical practices is a perpetual challenge for modern medicine. In this regard, computational intelligence has emerged as a promising subject for developing smart systems in healthcare practice [1]. Computational intelligence allows implementing automatic tools, enabling physicians to provide patients with a better quality of attention by performing early and accurate diagnosis and improving treatment. Furthermore, automatic systems and technologies based on computational intelligence have proven to be useful solutions to be applied in clinical practice. Some important advantages of intelligent automatic methods over traditional ones include better efficiency, accuracy, consistency, more time available for face-to-face consultation, and more time for critical tasks and critical cases, among others [2].

A specific subject where the learning capabilities of computational intelligence methods is very helpful to improve medical practice is analyzing and processing electronic medical records (EMRs). EMRs refer to digital records, collected by the individual medical practice, that contain the general health information of patients [3]. They usually consist of several types of health data, including, but not limited to, demographics, medical family history, medication, allergies, test results, and radiology images.

Currently, the majority of medical history recording products are based on predefined templates, which provide very limited freedom for writing patients medical records. Structured data entry is a hindrance to the usability of medical record applications, and is frowned upon by physicians who usually prefer to document using free text [4]. In addition, structured data entry systems do not take into account the particularities of the annotations of each physician, failing to effectively record the singularities of medical consultations. Alerts and suggestions offered by conventional products are generally based on previously defined rules, or according to mechanisms whose behavior remains the same throughout its operational life. The dissatisfaction of physicians with actual medical history recording products is increasing as they gain knowledge about automatic assistant tools. Consequently, physicians are increasingly aspiring to have sophisticated tools that help facilitate their clinical practice during medical consultations.

The research reported in this article is motivated by the need to further explore new ways of capturing, storing, and fostering medical reasoning. Thus, a formal proposal must be conceived to provide an accurate tool capable of following medical reasoning, aiming at helping physicians during medical consultations. In this line of work, this article presents a novel approach to represent clinical knowledge, which supports an appropriate methodology to follow reasoning in medical consultation. Likewise, the proposed representation does not pose formal restrictions to physicians, as they usually find when using common clinical data entry systems. An instance-based learning method is also introduced to provide suggestions in order to help during the process of registering a medical consultation. The developed system extends Praxis [5], a software used to follow medical reasoning with no templates, based on the accumulation of case types used to provide suggestions for subsequent cases.

The proposed approach was evaluated for a case study in which more than 50 advanced medical students had collaborated. Students tested the feasibility of the approach by using a proof-of-concept prototype. The performance of the proposed learning method was found to be satisfactory after being evaluated on 250 real instances constructed by the students. Results showed that the learning method was able to produce suggestions in a reasonable time frame, even when processing large volumes of data. The results suggest that the proposed approach was useful to accelerate the process of taking notes, since a convergence towards a high speed of completed medical records was observed. A high potential impact on clinical care may be projected, considering that the results showed that the proposed approach was appropriate to follow physician reasoning, especially during medical consultations. As a benchmark, 62% of the students were able to speed up writing time during medical consultations.

The main contributions of the research reported in this article include: (i) a formal structure to accurately represent clinical knowledge, and support the main flows of medical consultations; (ii) an instance-based learning method able to help reduce the time it takes to write notes; and (iii) a novel tool to help meet healthcare goals, which reminds physicians to record essential data to fulfilling care goals.

The article is structured as follows. Section 2 introduces learning models for assistance in medical consultation. A review of related work on learning models for assisting medical professionals is presented in Section 3. Section 4 describes a model proposed for representing clinical knowledge and patient history. Several flows to address relevant scenarios of medical consultations are presented in Section 5. The main implementation details of the proposed instance-based learning method are described in Section 6. Sample results from the evaluation are presented in Section 7. Section 8 discusses the usability of the proposed method and main strategies to improve the results and reduce uncertainties. Finally, Section 9 presents the main conclusions of the research.

2. Learning Models for Assistance in Medical Consultation

Despite the fact that physicians are becoming increasingly familiar with electronic medical records, they continue to have difficulties in dealing with long lists of pre-conceived variables, usually included in EMR systems. Although conventional EMR systems are useful to achieve legible, accessible, and complete documentation of medical consultations, they are causing several difficulties for physicians who adopt them. In many cases, physicians spend a lot of time searching for an option that allows them to record what they really want to write. Unfortunately, conventional EMR systems are template-based products that generate poor quality data, due to long search mechanisms and excessive mandatory fields, which often add noise to the relevant patient information [4]. Worse, the time required to enter clinical information sometimes exceeds the time required to write it on paper. The rigid structure of the templates to be filled-in during medical consultations does not fit the reasoning of physicians, nor their way of thinking.

Improvements in medical consultation assistance could be achieved by taking advantage of systems that allow better management of clinical information. To achieve better assistance, physicians should be provided with new healthcare tools, considering that healthcare assistance during medical consultations is improved when the physician is able to:

(i) Efficiently record all the information of a medical consultation, by reducing the time spent on mere data entry in order to gain more time to interact with the patient.

(ii) Use automatic clinical suggestions to reach an accurate diagnostics, or an appropriate indication of treatments.

(iii) Reduce medical errors, resulting from the human condition of the professional.

(iv) Record each medical consultation considering the special relevance of the interoperability of clinical information.

(v) Reuse recorded information for statistical and research purposes.

Computational intelligence can be applied to solve the deficiencies of current EMRs. Machine learning methods can be used to learn features from previous registered healthcare data sets, in order to provide suggestions for diagnoses and treatments based on information previously registered. By applying computational intelligence, systems can automatically identify solutions of similar clinical cases and can subsequently incorporate the knowledge gained to assist physicians during medical consultations. Learning methods can also contribute to reduce error-prone steps during the sequence of clinical tasks and decisions. Inevitable errors of human-based clinical practice may be reduced, such as drug contraindications, medication allergies, adverse drug reactions, and forgetting recurrent aspects of chronic patients. Furthermore, machine learning methods can progressively enhance their accuracy based on feedback provided by their own use.

An effective medical informatics support system must be adapted to the real health environment. In addition, a clinical evaluation of the usefulness of the system in real clinical work should be considered to determine its real capacity during clinical practice.

3. Related Works

This section reviews related works regarding learning models that assist medical professionals during their clinical activities.

Decision support systems can detect patterns, provide recommendations, and predict future behaviors for clinical practice. Wang et al. [6] proposed an Intelligent Self-Learning EMR (ISLEMR) system used to generate treatment recommendations based on learning and patient similarity. ISLEMR considers a group of ad hoc similarity metrics, considering patient diagnoses, demographic data, vital signs, structured lab test results, and information from external systems. The patient information is used to present an ordered menu with inferred recommendations for treatment plans. The system was validated on a real case study in Beijing, China in 2014, considering data from twelve-thousand patients. Precision results up to 80% were achieved for the first 10 items of the recommended menu; however, the applied learning algorithm only considered structured data, which implies less precision in determining similarities of clinical cases. Klann et al. [7] proposed a learning approach based on Bayesian networks (BN) to generate adaptive and context-specific treatment menus from past clinical information of patients. Each menu recommends a starting point for physicians, suggesting an initial draft to treat a specific situation. The BN models the probabilistic relationships among orders and diagnoses, covering typical scenarios from different aspects of medicine. The system was evaluated on a hospital simulation, demonstrating accurate predictive capabilities and outperforming a similar association rule mining approach, especially over less frequent cases. Support vector machines (SVM) have also been applied as learning models for medical assistance. Nakai et al. [8] applied SVM to predict clinical practices to be prescribed by using the information from previous practices of the same patient. The validation over real data from the Japanese system for medical billing proved the high precision of the model when facing frequent clinical cases; however, low precision results were obtained when dealing with less common cases. Barbantan et al. [9] proposed a medical decision support system using SVM and natural language processing to discover relations between medical concepts.

The model was successfully used to identify relations between medical concepts to help diagnoses, medication predictions, and to detect health patterns in Boston, USA. Shen et al. [10] proposed a multi-agent case-based reasoning approach for clinical decisions. The system searches clinical cases by identifying important words and terminologies, whereas medication allergies, adverse drug reactions, coexisting diseases, and other complications are evaluated to discard candidate cases. The system achieved a 78% matching rate for illnesses with simple syndromes. Installé et al. [11] developed a clinical data miner software framework for supporting clinical diagnostic using electronic case report forms (eCRF) based on templates and spreadsheets. Machine learning techniques are applied over the information gathered by the eCRF. A survey indicated that the system was considered user-friendly, and all physicians approved the possibility of using it in their own future works. Zieba [12] proposed a service-oriented support decision system for the diagnose of medical problems using web services with learning capabilities applying SVM. The system was evaluated using ontological datasets and it was able to predict a diagnosis by generating decision rules with acceptable accuracy values. Benmimoune et al. [13] designed a hybrid medical platform to assist physicians during their clinical reasoning process using rule-based reasoning (RBR) for general clinical cases and case-based reasoning (CBR) for clinical experiences. The proposed platform gathers relevant information about the patient status using an adaptive questionnaire and searches for the most similar stored case, following the CBR approach. If no similar case is found, the platform applies an RBR approach to deduce a solution according to rules defined by medical experts. Neither the implementation nor the prototype of the proposed system was described. Wilk et al. [14] proposed a framework to assist patients with multi-morbidity conditions, considering patient preferences for suggesting customized clinical practice guidelines. Clinical guidelines are modeled using actionable graphs and first-order logic, and a secondary medical knowledge component is used to identify adverse interactions resulting from conflicting therapies. A high-level proof of concept implementation was presented to show the feasibility of the proposed framework but no real evaluation was proposed.

Praxis is an electronic medical records application, developed to streamline the entry of clinical data and improve medical practice [5]. It emulates the processes that physicians follow when they are recording clinical information. The software uses previously entered information to offer recommendations for registering a new consultation, according to the past practice of the physician user (i.e., suggesting a set of cases similar to the one being evaluated). Praxis applies an empirical approach and has been gradually improved over more than twenty-five years, to fit the North American medical system. Praxis does not apply computational intelligence to build an expert system for the recommendation of diagnoses and treatments.

A summary of related works reviewed in this section is presented in Table 1, reporting for each article the methods applied, the most relevant features of each research, and any identified weaknesses.

The analysis of related works allowed identifying several proposals applying computational intelligence and other learning-based methods for diverse health scenarios. Most existing systems focuses on providing suggestions for treatments and diagnoses, based on similarity metrics regarding relevant information from past medical assistance. Reviewed works are able to identify similar clinical cases in order to provide suggestions for diagnoses, prognosis, and treatments. Furthermore, they contribute to reducing error-prone steps during the clinical process. The system presented in this article contributes to this line of research, including specific differences with existing related works: it supports non-structured free text information to be used in the learning process, instead of just structured information [6]; a more effective learning approach is applied, which outperforms a Bayesian learning method such as the ones that have been previously used in the related literature [7]; suggestions are generated considering all similar case types (of different patients), instead of just previous information of the same patient [8]; and it does not rely on complex rules based on natural language processing, which limits the applicability of other suggestion systems [9].

Table 1. Summary of reviewed works.

Work	Method(s)	Relevant Features	Weaknesses
Wang et al. [6]	Ad hoc patient similarity algorithm.	Menu of inferred recommendations, real-time feedback.	Only considers structured data.
Klann et al. [7]	Bayesian networks.	Suggest initial drafts, reduce workload of physicians.	Relies on a small set of orders and diagnoses.
Nakai et al. [8]	Linear support vector machine.	Use information from previous practices, high precision for common cases.	Low precision when dealing with less common cases.
Barbantan et al. [9]	Natural language processing, support vector machine classifier.	Medical structured-related concept model, detect patterns about patient health.	Only evaluated on clinical phrases with more than one medical concept.
Shen et al. [10]	Language analysis, ad hoc matching.	Suggest diagnoses, prognosis and treatments.	Knowledge representation fails to analyze evolutionary contexts.
Installé et al. [11]	Preprocessing, machine learning techniques.	Reduce error-prone steps during diagnostics, user-friendliness.	Variable length array types not supported, not useful for longitudinal data capture.
Zieba [12]	Cost-sensitive support vector machine.	Web services with learning capabilities, generate decision rules.	Only acceptable accuracy values of decision rules.
Benmimoune et al. [13]	Rules for generic cases, case-based reasoning component.	Adaptive questionnaire according to patient profile.	No prototype was implemented.
Wilk et al. [14]	Actionable graphs, first-order logic.	Clinical guidelines for multi-morbidity conditions, considers patient preferences.	No real evaluation.

4. Clinical Knowledge Model to Follow Physician Reasoning

A formal model is proposed for representing clinical knowledge and patient history, including medical records.

4.1. Clinical Knowledge Base

A bottom-up modeling approach is used to present the proposed clinical knowledge model. Several entities are defined in order to specify a clinical knowledge base that describes information of real medical case types. All entities included in a clinical knowledge base are described in the following subsections.

4.1.1. Unit of Thought

As defined by Low [15], a unit of thought is a statement that describes a basic clinical idea. Let UT^M be a unit of thought registered by physician M. UT^M is denoted as $UT^M = <ptext, uqcn, uqpt, exph, terms, inuse, ctSchedule, M>$, where *ptext* denotes a string capable of containing structured or random data, *uqcn* indicates if the unit refers to information to be used only in a unique consultation, *uqpt* indicates if the unit refers to unique information of a specific patient, *exph* indicates if the unit contains exclusive data for physician use, *terms* detail associations with health terminological standards, *inuse* denotes if the unit is in use during a consultation, and *ctSchedule* indicates the frequency that a unit appears in a case type. A unit of thought used in a case type will reappear each time the case type is used, unless a specific frequency is defined by its *ctSchedule* attribute.

The set of all units of thought registered by physician M is denoted as UT_T^M. Let UT_1^M = $<ptext_1, uqcn_1, uqpt_1, exph_1, terms_1, inuse_1, ctSchedule_1, M>$ and $UT_2^M = <ptext_2, uqcn_2, uqpt_2, exph_2, terms_2, inuse_2, ctSchedule_2, M>$ be units of thought registered by physician M. A constraint on units of thought is defined in Equation (1), implying that each basic clinical idea is represented by a unique unit of thought.

Considering that text variations do not change the meaning of a basic clinical idea, an ad hoc function *equal* (defined in Equation (2)) is necessary to identify if two phrases represent the same clinical idea. The same clinical idea can be instanced containing both structured information and random data, which implies that two different text strings can represent the same clinical idea.

$$\left.\begin{array}{r} UT_1^M \in UT_T^M \\ UT_2^M \in UT_T^M \\ equal(ptext_1, ptext_2) \end{array}\right\} \Rightarrow UT_1^M = UT_2^M \tag{1}$$

$$equal(t_1, t_2) = \begin{cases} true & \text{if } t_1 \text{ and } t_2 \text{ describe} \\ & \text{the same clinical idea.} \\ false, & \text{otherwise.} \end{cases} \tag{2}$$

4.1.2. Conceptual Element

A conceptual element is composed of a set of units of thought grouped to represent a broader concept. Several attributes are used to model all possible features of a conceptual element. Let CE^M be a conceptual element registered by physician M. CE^M is denoted as $CE^M = <name, display, chron, setDesc>$, where *name* denotes the name of the element, *display* indicates the default display mode of its units of thought, *chron* indicates if the element refers to a chronic condition, and *setDesc* denotes a set of possible descriptors of the conceptual element. The set $setDesc = \{[desc_1, subset_1(UT_T^M)], ..., [desc_k, subset_k(UT_T^M)]\}$ is composed of several pairs, each one is used to model a possible option to describe a real condition of a conceptual element.

Two constraints are defined on conceptual elements. The constraint presented in Equation (3) implies that a conceptual element is identified by its *name*.

$$\left.\begin{array}{r} CE_1^M =< name_1, display_1, chron_1, setDesc_1 > \\ CE_2^M =< name_2, display_2, chron_2, setDesc_2 > \\ name_1 = name_2 \end{array}\right\} \Rightarrow CE_1^M = CE_2^M \tag{3}$$

The constraint presented in Equation (4) implies the uniqueness of each descriptor into a conceptual element. Several units of thought can be labeled under the same descriptor to define an identified sub set, describing a real condition of an element.

$$\left.\begin{array}{r} [desc_1, subset_1(UT_T^M)] \in setDesc \\ [desc_2, subset_2(UT_T^M)] \in setDesc \\ desc_1 = desc_2 \end{array}\right\} \Rightarrow subset_1(UT_T^M) = subset_2(UT_T^M) \tag{4}$$

4.1.3. Conceptual Component

A conceptual component is composed of a set of conceptual element references, grouped to define sections of clinical information. Each conceptual component represents a typical clinical data section, in which a physician generally groups the information of a medical consultation.

Let $CC^M = <id, secType, activeElems>$ be a conceptual component defined by physician M, identified by its *id* attribute. The *secType* attribute is used to represent the type of data section, such as physical examination, medicines, and laboratory indications. Each *secType* must belong to the *ALL-SECTION-TYPES* set, which models all possible sections of the patient medical records. The set *activeElems* = *{[elemName$_1$, activeDesc$_1$], ..., [elemName$_k$, activeDesc$_k$]}* indicates which descriptor is used for each conceptual element referenced in a conceptual component.

Two constraints are defined on the conceptual components domain. The constraint presented in Equation (5) implies that a conceptual component is identified by its *id* attribute.

$$\left.\begin{array}{l} CC_1^M =< id_1, secType_1, activeElems_1 > \\ CC_2^M =< id_2, secType_2, activeElems_2 > \\ id_1 = id_2 \end{array}\right\} \Rightarrow CC_1^M = CC_2^M \tag{5}$$

A second constraint presented in Equation (6) ensures the referential integrity of names and descriptors of the active elements, referenced from a conceptual component.

$$\left.\begin{array}{l} CC^M =< id, secType, activeElems > \\ [elemName, activeDesc] \in activeElems \end{array}\right\} \Rightarrow \begin{array}{l} \exists \text{ conceptual element } e =< name, ..., setDesc > / \\ e.name = elemName \wedge \exists\, d \in setDesc, \\ d.desc = activeDesc \end{array} \tag{6}$$

4.1.4. Case Type

Several conceptual components can be grouped by a unique name to label a complex scenario, representing a real case type that can occur during a physician's workday. Let CT^M be a case type registered by physician M. CT^M is denoted as CT^M=<name,{CC_1^M, ..., CC_n^M}, chron, chronicComponents>, where *name* indicates the name of the case type, the set *{CC_1^M, ..., CC_n^M}* describes a specific group of conceptual components, *chron* indicates if the case type is marked as chronic, and *chronComponents* denotes all components used to monitor chronic conditions.

Three constraints are defined on case types domain. The constraint presented in Equation (7) implies that a case type is identified by its *name*.

$$\left.\begin{array}{l} CT_1^M =< name_1, chron_1, comps_1, chronComponents_1 > \\ CT_2^M =< name_2, chron_2, comps_2, chronComponents_2 > \\ name_1 = name_2 \end{array}\right\} \Rightarrow CT_1^M = CT_2^M \tag{7}$$

The second constraint presented in Equation (8) implies that each conceptual component of a case type models a different section of the clinical information.

$$\left.\begin{array}{l} CT^M =< nc, \{CC_1^M, ..., CC_n^M\}, chron, chComps > \\ CC_i^M =< id_i, secType_i, subset_i) > \\ CC_j^M =< id_j, secType_j, subset_j) > \end{array}\right\} \Rightarrow \begin{array}{l} secType_i = secType_j \\ \Updownarrow \\ i{=}j\ \forall i, j \in \{1, n\} \end{array} \tag{8}$$

The third constraint presented in Equation (9) implies that each chronic conceptual component models a different section of chronic clinical information.

$$\left.\begin{array}{l} CT^M =< nc, comps, true, \{CC_{chron_1}^M, ..., CC_{chron_m}^M\} > \\ CC_{chron_i}^M =< id_i, secType_i, subset_i) > \\ CC_{chron_j}^M =< id_j, secType_j, subset_j) > \end{array}\right\} \Rightarrow \begin{array}{l} secType_i = secType_j \\ \Updownarrow \\ i{=}j\ \forall i, j \in \{1, m\} \end{array} \tag{9}$$

Finally, the clinical knowledge base (CKB) of a physician M is defined as $CKB^M = \bigcup_{i=1}^{n} CT_i^M$. i.e., the union of all case types registered by physician M.

4.2. Patient Representation

A data structure is used to organize the information of each patient, considering the most relevant groups of personal data sets. The proposed structure includes medical records of a patient's history, and it also considers the chronic information of each patient.

4.2.1. Patient Structure

Each patient is modeled as $P = <personalData, MR^P, chronicElems, chronicCaseTypes>$ where *personalData* denotes personal data (sush as patient and family background), MR^P denotes all medical records of the patient P, *chronicElems* indicates associations with chronic conceptual elements, and *chronicCaseTypes* indicates associations with chronic case types. The *chronicElems* set is defined as *chronicElems* = $\{[elemName_1, chronDesc_1], ..., [elemName_j, chronDesc_j]\}$, and it is used to remember the descriptors of the elements that describe the chronic conditions of a patient. Additionally, the set *chronicCaseTypes* = $\{caseTypeName_1, ..., caseTypeName_k\}$ is used to remember all chronic case types associated with a specific patient P.

4.2.2. Patient Medical Records

The set of medical records of a patient P is denoted by MR^P and contains all records included in the medical history of the patient. A medical record of patient P created at time t is denoted as mr_t^P and it is defined as $mr_t^P = <content,p,t>$, where *content* is a set of *[phrase, unit]* pairs, each one includes a unit of thought associated with a clinical phrase. Consequently, $MR^P = \{mr_{t_1}^P, mr_{t_2}^P, mr_{t_k}^P\}$ describes the history of a patient, containing k medical records.

Let $mr_t^P = <content,p,t>$ be a specific patient medical record, where *content* = $\{[phrase_1, unit_1], ..., [phrase_n, unit_n]\}$ is composed by one or more pairs of clinical information. A function *showRecord* is used to print the content of a medical record, taking into account all phrases included in the *content* of a medical record. Function *showRecord* only prints clinical phrases, no unit of thought is shown.

4.2.3. New Medical Record

Let $CKB_t^M = \{CT_1^M, CT_2^M, ..., CT_n^M\}$ be the composition of the clinical knowledge base of physician M at time t. A medical record mr_t^P is generated as a result of the interaction of physician M and patient P, during a consultation at time t.

Since a physician usually takes a case type CT_x^M as basis to record a specific consultation, a transformation T^* can be applied to generate a new medical record. Consequently, a record $mr_t^P = T^*(CT_x^M)$ is created, taking into account the active information of a selected case type. The active information of a case type is defined by the units of thought with *inuse* attribute in true. Transformation $T^*: CKB^M \rightarrow MR^P$ is defined as $T^*(CT) = mr$, where mr is generated by applying Algorithm 1.

Algorithm 1 New medical record of patient P

1: units $\leftarrow$ getAllUnitsIncludedIn(CT)

2: content $\leftarrow \varnothing$

3: **for** unit **in** units **do**

4:     **if** unit.inuse **then**

5:        **if** not (unit.uqpt or unit.exph) **then**

6:           itemCont $\leftarrow$ [copyCurrentText(unit.ptext), unit]

7:           content $\leftarrow$ content $\bigcup$ {itemCont}

8:        **end if**

9:     **end if**

10: **end for**

11: $\text{mr}_t^P \leftarrow$ <content,P,t>

Algorithm 1 starts by getting all units of thought referenced in a case type CT (line 1). The algorithm iterates over all referenced units to identify units of thought marked with *inuse* attribute (lines 3–4). Further, units marked with *uqpt* or *exph* attributes are not taken into account for creating a new medical record (line 5). A new data pair is created for each identified unit (line 6), each pair includes the identified unit of thought, and a copy of its current text presentation. All new pairs are joined to build the full content of the consultation record (line 7). Finally, mr_t^P is created as a new medical record, containing the full description of the consultation of patient P at time t.

5. Medical Consultation Flows

Different flows for address the most relevant scenarios that arise during medical consultations are presented. These scenarios describe usual situations of physician workday, including multiple diagnoses, and the attention of chronic patients.

5.1. Starting Attention of a Patient

Algorithm 2 details the first steps which occur during a medical consultations.

Algorithm 2 Start attention of patient P

1: showPersonalInfo(P.personalData)

2: showChronicElementDescriptors(P.chronicElems)

3: chronicCTs $\leftarrow$ getCaseTypesByNames(P.chronicCaseTypes)

4: **if** chronicCTs $\neq \varnothing$ **then**

5:     All case types included in chronicCTs are suggested to physician

6:     Physician select $\text{CT}^M_{chron_1}, ..., \text{CT}^M_{chron_k}$ to be used as basis

7:     CT^M_{merge} is build by merging $\text{CT}^M_{chron_1}, ..., \text{CT}^M_{chron_k}$ (Algorithm 8)

8:     applyCaseType(P, CT^M_{merge}) is called (Algorithm 3)

9: **end if**

10: Show message agents according its trigger conditions

11: Physician continues with patient attention

The physician starts the attention of patient P by opening a registry editor to record the information of the new medical consultation. Personal information is loaded (line 1) to introduce the patient. All descriptors of chronic elements (line 2) and all chronic case types (line 3) associated with the patient are presented and suggested to the physician, who can select the chronic case types that are appropriate to being applied into the consultation (lines 4–9). Before the physician continues with patient attention, all message agents are evaluated and shown according to its trigger conditions (line 10).

5.2. Selecting an Already Defined Case Type

The selection of an already defined case type allows the physician to efficiently reuse previously registered information. Algorithm 3 details how to apply a case type during a medical consultation.

Algorithm 3 starts by evaluating the *chronic* attribute of a case type CT^M (lines 1–2). If the case type is identified as chronic, a specific method for applying a chronic case is called (line 3). Otherwise, all elements referenced in the *components* attribute are determined, and its units of thought are marked as in use according *setUnitsInUse* auxiliary procedure. The auxiliary procedure encapsulates the logic of how units of thought are activated. The algorithm continues by showing all units marked as in use, and highlighting the units that are exclusive for physician use (lines 8–9). Finally, each message agent that has CT^M as a trigger condition is presented to the physician (line 10).

Procedure *setUnitsInUse* iterates over all conceptual elements of a case type (line 12). All units included in each conceptual element are identified (line 13), and each unit of thought is marked as in use according the values of its attributes (lines 14–25).

Algorithm 3 applyCaseType(P, CT^M)

1: CT^M = <name, components, chronic, chronicComponents> is selected

2: **if** chronic **then**

3:   applyChronicCaseType(P, CT^M) (Algorithm 4)

4: **else**

5:   elements ← getAllElementsIncludedIn(components)

6:   setUnitsInUse(elements,CT^M)

7: **end if**

8: Show all units with *isuse* attribute in true

9: Highlight all units with *exph* attribute in true

10: Show message agents that have CT^M as a trigger condition

11: procedure **setUnitsInUse**(elements,CT^M)

12: **for** element **in** elements **do**

13:   units ← getAllUnitsIncludedIn(elements)

14:   **for** unit **in** units **do**

15:     **switch** ()

16:     **case** unit.exph:

17:       unit.inuse = true

18:     **case** isTime(unit.ctSchedule, CT^M):

19:       unit.inuse = true

20:     **case** element.display ∧ isEmpty(unit.ctSchedule):

21:       unit.inuse = true

22:     **case** otherwise:

23:       unit.inuse = false

24:     **end switch**

25:   **end for**

26: **end for**

5.3. Chronic Patients Flow

A case type CT^M can be marked as a chronic case type CT^M_{chron} at any time. When a CT^M_{chron} is marked as chronic, its *chron* attribute is activated and its *chronicComponents* attribute is initialized with an empty set. The chronic components are defined the first time that the case type is used to monitor a chronic patient. Algorithm 4 details how a physician can apply a chronic case type CT^M_{chron}.

Algorithm 4 analyzes if it is the first time that a chronic case type CT^M_{chron} is used with a patient being evaluated (lines 1–2). In that case, elements referenced in usual conceptual components are determined, and its units of thought are activated by calling *setUnitsInUse* procedure (lines 3–4). If CT^M_{chron} was used in any previous consultation of the same patient (line 6), then its *chronic* components are taken into account each time the physician decides to apply the case type, since *chronic* components are used to monitor the evolution of a chronic condition. However, the first time that CT^M_{chron} is used to monitor the evolution of a patient, the physician needs to define all entities that they want to use as monitoring items (lines 7–12). In addition, it is mandatory that the physician specify the frequency of each new unit of thought, included in an element of a chronic component (lines 13–16). All entities defined in new chronic components are used to monitor the patient's chronic condition in subsequent consultations (line 17). Finally, the units of thought of the elements referenced in *chronic* components are marked as in use by applying *setUnitsInUse* procedure (lines 19–20).

Algorithm 4 applyChronicCaseType(P, CT^M_{chron})

1: A chronic case type CT^M_{chron} = <name, components, true, chronicComponents> is selected

2: **if** name $\notin$ P.chronicCaseTypes **then**

3:      elements $\leftarrow$ getAllElementsIncludedIn(components) $\triangleright$ First time of case type for patient P

4:      setUnitsInUse(elements,CT^M_{chron})

5: **else**

6:      **if** chronicComponents = $\varnothing$ **then**

7:          Evolution component $CC_{Evolution}$ emerges $\triangleright$ Chronic components defined by physician

8:          Physician defines all conceptual elements included in $CC_{Evolution}$

9:          CC_{others} can be defined to better monitor the chronic condition

10:         $CCs_{new} = CC_{Evolution} \cup CC_{others}$

11:         newMonitorElems $\leftarrow$ getAllElementsIncludedIn(CCs_{new})

12:         newChronUnits $\leftarrow$ getAllUnitsIncludedIn(newMonitorElems)

13:         **for** newChronUnit in newChronUnits **do**

14:            Physician needs to specify the frequency of newChronUnit

15:            newChronUnit.ctSchedule is updated

16:         **end for**

17:         chronicComponents $\leftarrow$ CCs_{new} the chronic case type is updated

18:      **end if**

19:      elements $\leftarrow$ getAllElementsIncludedIn(chronicComponents)

20:      setUnitsInUse(elements,CT^M_{chron})

21: **end if**

5.4. Usual Attention Flow

During an attention flow, a physician can take advantage of an already registered case type. Algorithm 5 shows how a case type can be used to record a frequent medical consultation scenario.

In Algorithm 5, a procedure waits until the physician selects a case type and applies it to the current consultation (lines 1–2). After a case type is applied, the physician can also make modifications in order to describe the accurate information of the entire clinical meeting (line 3). Each unit of thought marked as unique to the patient being evaluated is stored as personal data, and is removed from the current case type (lines 4–6). The algorithm continues by applying T^* transformation, which generates a new medical record for the patient's history (lines 7–8). All chronic conceptual elements used in the case type are associated with the patient. Furthermore, if the case type is chronic, it is associated as permanent patient data (lines 9–13). Each unit of thought marked as unique to the current consultation is removed before updating the CKB^M of the physician (line 14). To update CKB^M, the physician needs to specify if the current case type refers to a new workday scenario, or it is only an improvement over the previously selected case type (lines 14–21). Two data sets are modified after the usual attention flow: physician CKB^M and patient history, including an MR^P increment.

Algorithm 5 Usual attention flow of a patient P

1: $CT^M \leftarrow$ selectSimilarCT()

2: applyCaseType(P, CT^M) is called

3: Physician M define CT'^M by modifying the selected CT^M

4: personalInfo $\leftarrow$ getUqptUnits(CT'^M)

5: P.personalData.add(personalInfo)

6: $CT'^M \leftarrow$ removeUqptUnits(CT'^M) units marked with *uqpt* are removed

7: $mr_i^P \leftarrow T^*(CT'^M)$

8: $MR^P \leftarrow MR^P \bigcup \{mr_i^P\}$

9: chronElemts $\leftarrow$ getAllActiveChonicElementsInludedIn(CT'^M)

10: P.chronicElems.add(chronElemts)

11: **if** isChronic(CT'^M) **then**

12:     P.chronicCaseTypes.add(CT'^M.name)

13: **end if**

14: $CT'^M \leftarrow$ removeUqcnUnits(CT'^M) units marked with *uqcn* are removed

15: **if** CT'^M is saved as an improvement **then**

16:     $CT^M \leftarrow CT'^M$

17:     CKB^M is updated with the new version of CT^M

18: **else**

19:     $CT_{new}^M \leftarrow CT'^M$ is saved as a new case type

20:     $CKB^M \leftarrow CKB^M \bigcup \{CT_{new}^M\}$ the base is incremented

21: **end if**

5.5. New Case Type Flow

Algorithm 6 details the flow followed by the physician when they need to address a new case type that is not included in their CKB.

Since there is no case type to be re-used, Algorithm 6 needs to create an empty case type in which the new workday scenario can be detailed (lines 1–2). To define a new case type CT_{new}^M, the physician can re-use any predefined unit of thought, and can also create units of thought specifying new clinical phrases. Furthermore, predefined conceptual elements can be re-used and new elements can be created (lines 3–4). Each element defined by the physician is referenced from one clinical section. Therefore, new conceptual components are created in order to group elements sharing the same section type (lines 5–11). It is mandatory that the physician assigns a name to the new clinical case type. The case type can also be marked as chronic, and in that case, the physician needs to specify the chronic attribute of each new element, created while defining the new case type (lines 12–21). All units of thought marked as unique to the patient are stored as personal data, and are removed from CT^M_{new} (lines 22–24). Then, T^* transformation is applied to generate a new medical record in the patient's history (lines 25–26). All chronic conceptual elements of CT^M_{new} are associated with the patient, and if the case type is chronic it is associated as permanent patient data (lines 27–31). Finally, all units of thought marked as unique to current consultation are removed from the case type, and the clinical data base of the physician is enriched by including the new case type.

Algorithm 6 New case type flow for the attention of patient P

1: There is no CT^M selected by physician

2: $CT^M_{new} \leftarrow\; <$"new-name", $\varnothing$, false, $\varnothing >$ is created automatically

3: Physician creates new units UTs_{new} and new elements CEs_{new}

4: Physician defines sections, by using UTs_{new} and CEs_{new} or pre-defined

5: secTypes $\leftarrow$ *ALL-SECTION-TYPES*

6: newComponents $\leftarrow \varnothing$

7: **for** $secType_i$ **in** secTypes **do**

8:     $activeElems_i \leftarrow$ [elementName, activeDescriptor] pairs in $section_i$

9:     $CC_{new_i} \leftarrow\; <$ maxCCId() + 1, $secType_i$, $activeElems_i >$

10:     newComponents $\leftarrow$ newComponents $\bigcup \{CC_{new_i}\}$

11: **end for**

12: Physician assigns a unique name to attribute *name* of CT^M_{new}

13: Physician can mark CT^M_{new} as chronic

14: **if** CT^M_{new} is marked as chronic **then**

15:     **for** elem **in** CEs_{new} **do**

16:         Physician needs to specify the value of *elem.chron*

17:     **end for**

18:     $CT^M_{new} \leftarrow\; <$name, newComponents, true, $\varnothing >$

19: **else**

20:     $CT^M_{new} \leftarrow\; <$name, newComponents, false, $\varnothing >$

21: **end if**

22: personalInfo $\leftarrow$ getUqptUnits(CT^M_{new})

23: P.personalData.add(personalInfo)

24: $CT^M_{new} \leftarrow$ removeUqptUnits(CT^M_{new})

25: $mr^P_t \leftarrow T^*(CT^M_{new})$

26: $MR^P \leftarrow MR^P \bigcup \{mr^P_t\}$

27: chronElemts $\leftarrow$ getAllActiveChonicElementsInludedIn(CT^M_{new})

28: P.chronicElems.add(chronElemts)

29: **if** isChronic(CT^M_{new}) **then**

30:     P.chronicCaseTypes.add(CT^M_{new}.name)

31: **end if**

32: $CT^M_{new} \leftarrow$ removeUqcnUnits(CT^M_{new})

33: $CKB^M \leftarrow CKB^M \bigcup \{CT^M_{new}\}$ the base is incremented

---

5.6. Temporal Case Type Flow

By applying a temporal case type, the history of patient P is updated, and the MR^P set is incremented with a new patient medical record. However, there is no change in physician CKB^M. Algorithm 7 details the use of a temporal case type.

Algorithm 7 Temporal case type flow for the attention of patient P

1: $CT^M \leftarrow$ selectSimilarCT()

2: applyCaseType(P, CT^M) is called

3: Physician M define CT'^M by modifying the selected CT^M

4: Physician M marks CT'^M as a temporal case type

5: personalInfo $\leftarrow$ getUqptUnits(CT'^M)

6: P.personalData.add(personalInfo)

7: $CT'^M \leftarrow$ removeUqptUnits(CT'^M)

8: $mr_t^P \leftarrow T^*(CT'^M)$

9: $MR^P \leftarrow MR^P \bigcup \{mr_t^P\}$

10: chronElemts $\leftarrow$ getAllActiveChonicElementsInludedIn(CT'^M)

11: P.chronicElems.add(chronElemts)

12: CT'^M is deleted

Algorithm 7 is triggered after the physician assigns a temporal mark over an applied and modified case type (lines 1–4). As any other case type, all units marked as unique to the patient are stored as personal data, and are removed from the temporal case type (lines 5–7). T^* transformation is also applied to create a new medical record (lines 8–9). Each chronic conceptual element referenced in the temporal case is permanently associated with the patient being evaluated (lines 10–11). The temporal case type is finally removed, since it is marked to be not re-used (line 12).

5.7. Multiple Case Types Flow

A physician can use more than one case type as the basis during the same medical consultation. Several rules are used to combine all conceptual components of each case type involved. To combine conceptual components, their active conceptual elements are accurately merged. The merge process takes into account the active elements described in the usual components, and active elements described in chronic components. Algorithm 8 details the method used to merge different case types.

Algorithm 8 starts by identifying the conceptual components included in each case type (lines 1–4). An ad hoc *merge* function is used to combine all identified components (line 5). Function *merge* is also applied over chronic conceptual components (lines 6–8). The algorithm continues by creating a case type CT_{merge}^M, which includes all merged components (lines 9–14). Then, the case type is applied and can be modified by the physician (lines 15–17). Each unit of thought marked as unique to the patient is taken into account as usual, it is stored as personal data and removed from the case type (lines 18–20). Likewise, a new medical record is created by applying T^* transformation (lines 21–22). Each chronic conceptual element referenced in CT_{merge}^M is permanently associated with the patient being evaluated, as well as any original chronic case type (lines 23–30). Finally, the used case type is deleted after concluding the consultation (line 31).

Algorithm 8 Multiple case types flow for the attention of patient P

1: $CT_1^M \leftarrow$ selectSimilarCT()

2: $CT_2^M \leftarrow$ selectSimilarCT()

3: $comps_1 \leftarrow$ getAllComponents(CT_1^M)

4: $comps_2 \leftarrow$ getAllComponents(CT_2^M)

5: $comps_{merge} \leftarrow$ **merge**($comps_1$, $comps_2$)

6: $chronComps_1 \leftarrow$ getAllChronicComponents(CT_1^M)

7: $chronComps_2 \leftarrow$ getAllChronicComponents(CT_2^M)

8: $chronComps_{merge} \leftarrow$ **merge**($chronComps_1$, $chronComps_1$)

9: $name_{merge} \leftarrow$ concat(CT_1.name,CT_2.name)

10: **if** $chronComps_{merge} \neq \varnothing$ **then**

11:     $CT_{merge}^M = <name_{merge},comps_{merge}, \text{true}, chronComps_{merge} >$

12: **else**

13:     $CT_{merge}^M = <name_{merge},comps_{merge}, \text{false}, \varnothing >$

14: **end if**

15: CT_{merge}^M is auto-selected

16: applyCaseType(P, CT_{merge}^M) is called

17: Physician M define $CT_{merge}'^M$ by modifying CT_{merge}^M.

18: $personalInfo \leftarrow$ getUqptUnits($CT_{merge}'^M$)

19: P.personalData.add(personalInfo)

20: $CT_{merge}'^M \leftarrow$ removeUqptUnits($CT_{merge}'^M$)

21: $mr_t^P \leftarrow T^*(CT_{merge}'^M)$

22: $MR^P \leftarrow MR^P \cup \{mr_t^P\}$

23: $chronElemts \leftarrow$ getAllActiveChonicElementsInludedIn($CT_{merge}'^M$)

24: P.chronicElems.add(chronElemts)

25: **if** isChronic(CT_1^M) **then**

26:     P.chronicCaseTypes.add(CT_1^M.name)

27: **end if**

28: **if** isChronic(CT_2^M) **then**

29:     P.chronicCaseTypes.add(CT_2^M.name)

30: **end if**

31: $CT_{merge}'^M$ is deleted

---

6. Instance-Based Learning

A learning method is proposed in order to generate suggestions for physicians. The proposed method is based on an ad hoc similarity metric, designed to compare the similarity between clinical case types.

6.1. Instance-Based Learning Method

An instance-based learning method is designed with the aim to provide suggestions for physicians. The proposed method takes into account the clinical knowledge base of a physician, in order to present suggestions based on previously defined case types. A register editor where a physician can take advantage of the proposed instance-based learning method is also introduced.

6.1.1. Register Editor

The register editor is an interface in which a physician can register a consultation appointment. The register editor presents personal information of the patient being evaluated, and includes an area for writing all details of a medical consultation. The main features of the register editor are illustrated in Figure 1, including a list of case type suggestions.

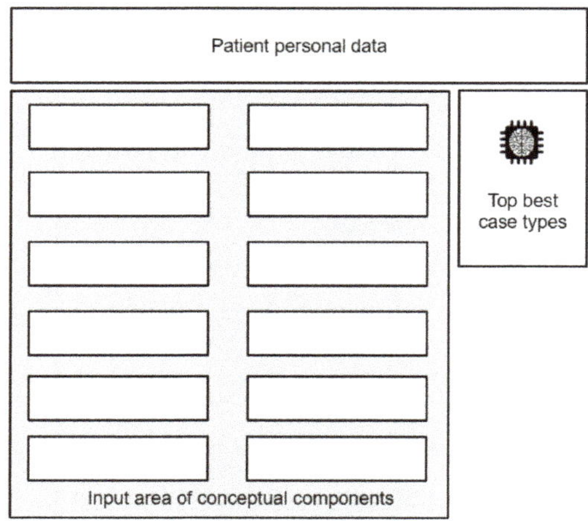

Figure 1. Main features of register editor.

The input area of the register editor is designed with the aim of registering a consultation in an organized structure, grouping information by clinical section types. When a physician writes in the input area, a case type $CT_{current}$ is automatically created, based on the information included in each section type. As a relevant feature, a list of similar case types is included in the register editor as suggestions for the physician. The suggested list is based on the top best values of a similarity metric, applied to compare the information of $CT_{current}$ against all case types previously registered.

6.1.2. Learning Method

A learning method is applied to determine the case types that best match with the clinical scenario of the patient being evaluated, according to an ad hoc similarity metric. A list of similar case types is suggested each time the physician modifies the information of the patient being evaluated. The list of similar case types is updated when introducing or removing any clinical phase during a medical consultation.

The proposed learning method implements a lazy approach [16], since the training stage of learning is delayed until a new case type draft must be evaluated. To evaluate a new case type draft, all previously defined case types are processed as training examples, and a similarity metric is applied to determine the most similar candidates. Algorithm 9 details how the instance-based learning method is implemented, seeking to suggest similar case types.

The learning method described in Algorithm 9 is triggered each time the physician modifies any aspect of the consultation being evaluated. An auxiliary case type $CT_{current}$ is created based on the information detailed by the physician in their register editor (line 1). Sentences without any meaningful word are not taken into account by the learning method (line 2). A step to remove duplicate units of thought is applied, since a physician could write duplicate clinical phrases in their register editor (line 3). Moreover, an array used to identify top best *similarity* metrics is initialized with empty values (line 4). The algorithm continues by iterating over all case types included in the physician clinical knowledge base (line 5). For each iteration, the similarity between $CT_{current}$ and any other case type is calculated in order to update the array of top best metrics (lines 6–9). Finally, the case types with top best metrics are returned as suggestions to the physician (lines 11–12).

Algorithm 9 Instance-based learning method

1: $CT_{current} \leftarrow$ new CT(registerEditor.content)

2: $CT_{current} \leftarrow$ removeNonMeaningfulUnits($CT_{current}$)

3: $CT_{current} \leftarrow$ removeDuplicateUnits($CT_{current}$)

4: topBest $\leftarrow$ Initialize array with t empty values

5: **for** CT_i in CKB^M **do**

6:     similarityMetric $\leftarrow$ *similarity*($CT_{current}$,CT_i)

7:     **if** similarityMetric is better that worstSimilarity(topBest) **then**

8:         topBest $\leftarrow$ replaceWorst(topBest,CT_i)

9:     **end if**

10: **end for**

11: topBest $\leftarrow$ removeEmptyValues(topBest)

12: **return topBest**

6.1.3. Using Suggested Case Types

A physician can select a suggested case type as a basis of writing a medical consultation. After a case type is applied, the input area of the register editor is updated by using the information defined in the selected case type. By using suggested case types, previously registered phases are re-used and the time spent writing the details of a consultation is reduced. Suggested case types can also remind physicians to verify important clinical aspects of their patients. Moreover, taking advantage of previously written sentences is useful when physicians need to address recurrent aspects of chronic patients.

6.2. Similarity Metric

A similarity metric is introduced in order to compare two case types of a clinical knowledge base. The proposed metric takes into account the similarity between conceptual components of different case types. Consequently, the similarity value between two case types is determined by the weighted similarities of their conceptual components.

6.2.1. Similarity Metric Definition

Sadegh-Zadeh [17] introduced the concept of *diagnostic relevance*, which applies fuzzy logic to evaluate the relevance of causal events associated with a clinical diagnosis. The proposed method is based on a similar idea, where the concept of *medical relevance* is considered to evaluate the relevance of conceptual components associated with a clinical case type.

Let $comp_{i,sec}$ be a conceptual component of case type CT_i defined with sec section type, where sec belongs to ALL-SECTION-TYPES. The set ALL-SECTION-TYPES is introduced in Section 4 as a set that defines all possible clinical sections of the patient medical records. The similarity between two case types is denoted as *similarity*, and is defined by Equation (10).

$$similarity(CT_1, CT_2) = \sum_{sec} w_{sec} \times similarityCC_{sec}(comp_{1,sec}, comp_{2,sec}) \qquad (10)$$

In Equation (10), each w_{sec} defines the weight of a component with sec section type. Consequently, the conceptual components of case types influence the *similarity* metric according to their sec section type. The similarity weight of a clinical section type is determined by its medical relevance. The medical relevance is used to define the weight of each section type belonging to ALL-SECTION-TYPES = {sec_1, ..., sec_n}, as $W_{sec} = medRelevance(sec) / \sum_{sec_1}^{sec_n} medRelevance(sec_i)$.

The medical relevance of clinical section types must be defined based on background knowledge of the health area. Accurate weights of conceptual components provide a mechanism for reducing the impact of irrelevant features in the similarity metric [16]. As an example, health background knowledge suggests that the section for excuse notes should weigh less than the diagnosis section.

The following subsection presents the *similarityCC* function, introduced in Equation (10) for comparing two conceptual components of different case types.

6.2.2. Similarity between Components

To compare conceptual components, the similarity metric takes into account the units of thought included in all elements of conceptual components. The similarity between conceptual components is defined by Equation (11), which is aimed at comparing components sharing the same section type *sec*.

$$similarityCC_{sec}(cc_1, cc_2) = \begin{cases} 0, & \text{if } cc_1.secType \neq sec \vee cc_2.secType \neq sec \\ includedUTs(units(cc_1), units(cc_2)), & \text{otherwise} \end{cases} \tag{11}$$

Function *includedUTs*: $UT \times UT \rightarrow [-1,1]$ is applied to compare two sets of units of thought. Equation (12) presents *includedUTs* by considering different scenarios.

$$includedUTs(units_1, units_2) = \begin{cases} 0, & \text{if } units_1 = \emptyset \\ -1, & \text{if } units_1 \neq \emptyset \wedge units_2 = \emptyset \\ \max\{\sum_{u_1 \in units_1} \frac{belongs(u_1, units_2)}{|units_2|}, -1\}, & \text{otherwise} \end{cases} \tag{12}$$

If the first parameter $units_1$ of function *includedUTs* is an empty set, there is no unit of thought that can contribute as similarity data, then zero value is returned. Another exceptional case occurs when $units_2$ does not describe any information. If the second parameter is an empty set, the worst value of similarity must be returned because $units_1$ details clinical data not considered by $units_2$. A complex scenario arises when function *includedUTs* evaluates non-empty parameters. In that case, each unit of $units_1$ is analyzed in order to evaluate its inclusion into $units_2$, and a positive weight is determined for units that belong to both sets. In addition, a limit of maximum deference could be applied if $units_1$ and $units_2$ are significantly different and $units_1$ is bigger than $units_2$.

An auxiliary function *belongs(unit, units)* presented by Equation (13) is required to determine if a specific unit of thought belongs to a set of units. A *unit* that contradict the ideas represented by the *units* set is negatively weighted.

$$belongs(unit, units) = \begin{cases} 1, & \text{if } \exists\, u_{same} \in units \,/\, equal(unit, u_{same}) \\ -1, & \text{otherwise.} \end{cases} \tag{13}$$

6.2.3. Similarity Metric Algorithm

The metric detailed in Algorithm 10 calculates the similarity of a case type $CT_{current}$ regarding any other case type. To achieve the final value of the similarity metric, Algorithm 10 needs to calculate similarity values of several conceptual components.

Algorithm 10 starts by initializing the similarity metric with a neutral value (line 2). Then, all section types of conceptual components included in analyzed case types are identified (line 3). After identifying the section types that influence the similarity metric, a relative weight factor is determined in order to accurately weigh the influence of each identified section type (line 4). Each section type with a positive weight of similarity is taken into account to calculate the value of the metric (lines 5–6).

Algorithm 10 Similarity metric

1: function ***similarity***($\text{CT}_{current}$, CT_i)
2: $\quad$ similarity $\leftarrow 0$
3: $\quad$ involvedSecTypes $\leftarrow$ sectionTypesOf($\text{CT}_{current}$) $\bigcup$ sectionTypesOf(CT_i)
4: $\quad$ relativeWeight $\leftarrow$ relativeWeightFactor(involvedSecTypes)
5: $\quad$ **for** sec **in** involvedSecTypes **do**
6: $\quad\quad$ **if** sec.weight $\neq 0$ **then**
7: $\quad\quad\quad$ cachedSimilarityCC $\leftarrow$ getSimilarityCCValueFromCache(sec, CT_i)
8: $\quad\quad\quad$ **if** cachedSimilarityCC is hitted **then**
9: $\quad\quad\quad\quad$ similarityCC $\leftarrow$ cachedSimilarityCC
10: $\quad\quad\quad$ **else**
11: $\quad\quad\quad\quad$ units$_{current}$ $\leftarrow$ getUnitsBySectionType(sec,$\text{CT}_{current}$)
12: $\quad\quad\quad\quad$ units$_i$ $\leftarrow$ getUnitsBySectionType(sec,CT_i)
13: $\quad\quad\quad\quad$ **if** units$_{current}$ $\neq \varnothing$ **then**
14: $\quad\quad\quad\quad\quad$ **if** units$_i$ $\neq \varnothing$ **then**
15: $\quad\quad\quad\quad\quad\quad$ includedUTs $\leftarrow 0$
16: $\quad\quad\quad\quad\quad\quad$ **for** unit$_{current}$ **in** units$_{current}$ **do**
17: $\quad\quad\quad\quad\quad\quad\quad$ belongs $\leftarrow$ *belongs*(unit$_{current}$, units$_i$)
18: $\quad\quad\quad\quad\quad\quad\quad$ **if** belongs **then**
19: $\quad\quad\quad\quad\quad\quad\quad\quad$ includedUTs $\leftarrow$ includedUTs $+ \frac{1}{|units_i|}$
20: $\quad\quad\quad\quad\quad\quad\quad$ **else**
21: $\quad\quad\quad\quad\quad\quad\quad\quad$ includedUTs $\leftarrow$ includedUTs $- \frac{1}{|units_i|}$
22: $\quad\quad\quad\quad\quad\quad\quad$ **end if**
23: $\quad\quad\quad\quad\quad\quad$ **end for**
24: $\quad\quad\quad\quad\quad\quad$ similarityCC $\leftarrow$ max\{includedUTs, -1\}
25: $\quad\quad\quad\quad\quad$ **else**
26: $\quad\quad\quad\quad\quad\quad$ similarityCC $\leftarrow -1$
27: $\quad\quad\quad\quad\quad$ **end if**
28: $\quad\quad\quad\quad$ **else**
29: $\quad\quad\quad\quad\quad$ similarityCC $\leftarrow 0$
30: $\quad\quad\quad\quad$ **end if**
31: $\quad\quad\quad\quad$ putSimilarityCCValueInCache(similarityCC,sec,CT_i)
32: $\quad\quad\quad$ **end if**
33: $\quad\quad\quad$ similarity $\leftarrow$ similarity + (relativeWeight * similarityCC)
34: $\quad\quad\quad$ bestRemain $\leftarrow$ upperBound(sec, involvedSecTypes)
35: $\quad\quad\quad$ **if** similarity + bestRemain $<$ worstSimilarity(*topBest*) **then**
36: $\quad\quad\quad\quad$ *invalidateSimilarityCCValuesOnCache(CT_i)*
37: $\quad\quad\quad\quad$ **throw** *discard-low-similarity*
38: $\quad\quad\quad$ **end if**
39: $\quad\quad$ **end if**
40: $\quad$ **end for**
41: $\quad$ **return** similarity

For each identified section type, the similarity of components sharing the same section type must be calculated. A cache containing values of similarity between conceptual components is used to improve the performance of the proposed metric (lines 7–9). To calculate the similarity between conceptual components, the sets of units of thought included in each component are determined. Algorithm 10 implements the rules introduced by Equation (12) for calculating the similarity between two sets of units of thought, including the general scenario (lines 15–24) for non-empty sets, and exceptional scenarios (line 26 and line 29) to address singular situations of empty sets. Moreover, the calculated value of component similarity is cached, to be used in the future (line 31). The partial value of the similarity metric is updated after calculating the similarity between each pair of conceptual components. For each pair of components, the partial similarity is affected by the similarity of the components according to a relative weight factor (line 33).

To detect low values of similarity, an upper bound is calculated in order to determine a maximum possible value of similarity (line 34). If the similarity between two case types is detected early as low, it is not required to calculate its exact value. All case types with low *similarity* are discarded early, and their partial values of component similarity are removed from the cache as they are not fully calculated (lines 35–37). At last, a final value of similarity is returned after iterating over all involved section types (line 41).

6.3. Implementation of Similarity Metric

The similarity metric is an essential feature of the proposed learning method. The metric must be able to accurately compare the similarity between clinical case types, and it also needs to execute as quickly as possible. The similarity metric is highly demanded in virtue of the lazy approach of the learning method. Therefore, several techniques of indexing and cache are applied for reducing the metric execution time. All optimizations implemented to reduce the execution time of the *similarity* metric are presented in the following paragraphs.

Compare units by canonical form. The operator *equal* for units of thought is used to determine if two different sentences represent the same clinical idea. To implement the *equal* operator, a canonical transformation is applied over the units being compared. Two transformations are applied by comparing a pair of units of thought. For each unit of thought, structured information and random data are removed, in order to achieve the canonical form. Finally, a raw string comparison between both canonical forms is evaluated. Original units of thought are identified as equal only if they coincide in their canonical form.

Zero similarity value. Function *similarity(CT_1, CT_2)* is called to calculate the similarity between a case type CT_1 and another case type CT_2. Both case types are composed by conceptual components that influence the similarity metric according to its section type weights. However, an empty component of CT_1 cannot provide similarity information since it does not have associations with units of thought. If a conceptual component of CT_1 is empty, its similarity in regard to any other component is zero. No calculation is performed over the empty components of CT_1, rather all computational effort is performed over its non-empty components.

Comparing with empty components. All components of a case type CT_1 are analyzed when calculating the similarity of CT_1 in regard to another case type CT_2. Each conceptual component of CT_1 should be compared against a component of CT_2 with the same section type. If CT_2 does not include a conceptual component with the same section type, then a value representing the biggest difference of similarity is returned without performing additional calculations.

Cache of previous similarity values. The proposed similarity metric provides a mechanism for comparing different case types. However, the metric is not based on case types themselves, but on their conceptual components. Due to the high need of obtaining similarities between conceptual components, a cache is designed for containing pre-calculated values of component similarities. Figure 2 shows the structure used to maintain recent values of similarities, and how similarity values are cached for each case type included in the clinical knowledge base of a physician. The proposed structure is able to cache the last value of similarity of all conceptual components of each case type.

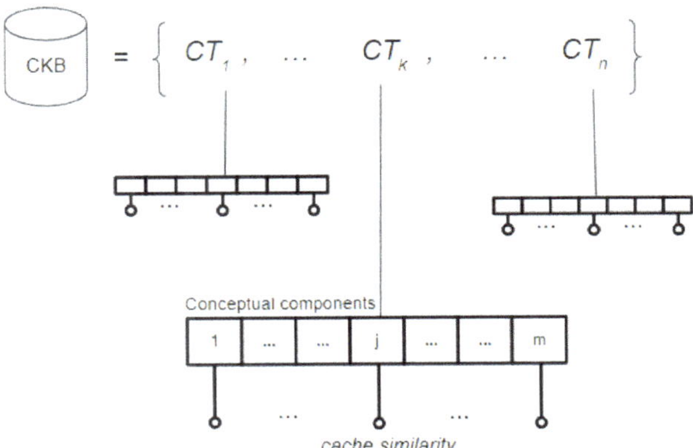

Figure 2. Cache structure for similarity between components. The clinical knowledge base is composed of case types, each one containing similarity cached values of its conceptual components.

After evaluating the similarity between a specific case type in regard to any other case type, all values of component similarities are stored in the cache. Figure 3 introduces a scenario in which a "Case type A" is slightly modified, by only changing the information described in one of its conceptual components. Several highlighted values of component similarities are obtained from the cache. Furthermore, a high cache hit ratio should be achieved after re-using any other case type and applying a few modifications.

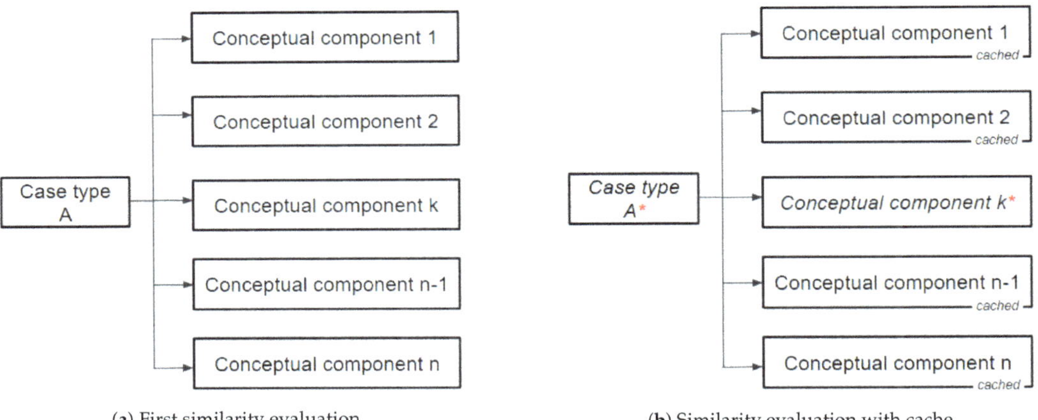

(**a**) First similarity evaluation

(**b**) Similarity evaluation with cache

Figure 3. Use of similarity cache values.

Discard non-promising candidates. The proposed learning method is designed to suggest the best case types that can be applied during a medical consultation. Top best case types are identified according to best *similarity* metric values, and only t best case types are presented to the physician.

The similarity function separates the first k section types from the rest of the ALL-SECTION-TYPES set, as described in Equation (14).

$$similarity(CT_1, CT_2) = \sum_{sec=sec_1}^{sec_N} w_{sec} \times similarityCC_{sec}$$

$$= \sum_{sec=sec_1}^{sec_k} w_{sec} \times similarityCC_{sec} + \sum_{sec=sec_k+1}^{sec_N} w_{sec} \times similarityCC_{sec} \tag{14}$$

$$\leq \sum_{sec=sec_1}^{sec_k} w_{sec} \times similarityCC_{sec} + \sum_{sec=sec_k+1}^{sec_N} w_{sec} = \sum_{sec=sec_1}^{sec_k} w_{sec} \times similarityCC_{sec} + R_{constant}^{k+1}$$

Equation (15) presents an upper bound inferred by simplifying (14), which can be used for discarding case types with low similarity values.

$$similarity(CT_1, CT_2) \leq \sum_{sec=sec_1}^{sec_k} w_{sec} \times similarityCC_{sec} + R_{constant}^{k+1} \tag{15}$$

Several component similarity values (*similarityCCs*) need to be computed to obtain the final value of *similarity(CT$_1$, CT$_2$)*. An upper bound is identified after determining the value of *similarityCC$_{sec_k}$*. After calculating the similarity of the *k*th conceptual component, it is possible to use an upper bound to discard a case type with a low similarity value. Each case type whose upper bound of similarity is lower than the worst element of the top best metrics is considered a non-promising candidate, and no more computational effort is expended to calculate its final similarity value.

7. Experimental Validation

This section presents the experimental validation of the proposed approach on a real case study, which served as a basis for evaluating the practical aspect of this research.

7.1. Problem Instances

The source *Clinical cases in primary care* [18] was used for evaluating the proposed approach. The source is a multi-authored publication that covers a wide range of clinical scenarios of primary care.

7.1.1. Prerequisites for Building Case Type Instances

The collaboration of advanced medical students was requested with the intention of registering as many clinical scenarios as possible. Students were instructed to record the primary care scenarios described in *Clinical cases in primary care* as new clinical CTs. In order to group all the information recorded, it was necessary to implement procedures for exchanging clinical CTs. The *export* and *import* procedures were used to exchange different CTs.

A procedure to *export* a given CT was implemented. The procedure extracts a CT from a specific clinical knowledge base CKB, and it also anonymizes any information that refers to the person who wrote (owner) the CT. The *import* procedure consolidates the information of a specific CT into a target CKB. A new CT is inserted into the target CKB, replacing any anonymized reference of the original owner with the person who owns the target CKB. Importing a CT is a complex procedure, which must avoid the generation of duplicate units of thought, and has to merge the conceptual elements of the new CT with those existing in the target CKB.

7.1.2. Building Case Type Instances

The set of clinical cases specified in the publication *Clinical cases in primary care* was distributed to be evaluated by 50 advanced medical students. Each student had to evaluate three different cases, and each clinical case was assigned to at least one student. Furthermore, each student was instructed to contribute two additional clinical cases, defined as variants of those presented in the clinical source.

All scenarios of primary care detailed in the clinical source were successfully registered by the group of students, including variants of repeated clinical scenarios. Algorithm 11 details how a single CKB was loaded with 250 scenarios of primary care, based on information registered by students.

Algorithm 11 Building case types

1: **for** i = 1 to length(STUDENT-LIST) **do**
2:     $student_i \leftarrow$ STUDENT-LIST[i]
3:     $CKB_{student_i} \leftarrow \varnothing$
4:     **for** j = 1 **to** 3 **do**
5:         $k\text{-}index \leftarrow mod(3(i - 1) + j, length(\text{CASE-SOURCE}))$
6:         $CT_{i_j} \leftarrow student_i$ records case number $k\text{-}index$ of CASE-SOURCE
7:         $CKB_{student_i} \leftarrow CKB_{student_i} \cup \{CT_{i_j}\}$
8:     **end for**
9:     $CT_{i_{v_1}} \leftarrow$ first variant of case type included in $CKB_{student_i}$
10:     $CKB_{student_i} \leftarrow CKB_{student_i} \cup \{CT_{i_{v_1}}\}$
11:     $CT_{i_{v_2}} \leftarrow$ second variant of case type included in $CKB_{student_i}$
12:     $CKB_{student_i} \leftarrow CKB_{student_i} \cup \{CT_{i_{v_2}}\}$
13: **end for**
14: $CKB_{all} \leftarrow \varnothing$
15: **for** i = 1 to length(STUDENT-LIST) **do**
16:     $student_i \leftarrow$ STUDENT-LIST[i]
17:     **for** j = 1 **to** 5 **do**
18:         $CT_{i_{jexported}} \leftarrow export(j, CKB_{student_i})$
19:         $import(CT_{i_{jexported}}, CKB_{all})$
20:     **end for**
21: **end for**
22: **return** CKB_{all}

Algorithm 11 starts by initializing all CKBs of the students (STUDENT-LIST) selected for recording new CTs (lines 1–3). Each student is expected to treat three fictitious patients suffering from one of the specific conditions of the clinical source (lines 5–6). Moreover, two variants contributed by each student are also registered (line 9 and line 11). Therefore, the CKB of each student is enriched with five new CTs (line 7, line 10, and line 12). The algorithm continues by initializing a single CKB_{all} that groups all information recorded by all students (line 14). Each registered CT is exported using the *export* procedure, and the *import* procedure is applied to consolidate the exported CT into the CKB_{all} (lines 15–21). Finally, the CKB_{all} which contains all the 250 registered CTs (five contributed by each of the 50 students) is returned (line 22).

7.2. Parameter Settings of Similarity Weight

For the purposes of the experimental evaluation, the set of clinical section types was defined following the Uruguayan health model. The set of clinical section types was defined as *ALL-SECTION-TYPES = {Diagnosis, Consultation reason, Current illness, Physical examination, Medication, Studies, Procedures, Referral, Message agents, Advisors, Excuse notes, Observations}*.

The similarity weight of a clinical section type is given by its medical relevance. A simple medical relevance criteria was applied to give greater weight to the most important section types. Four levels of importance were defined in order to consider qualitative ranges of medical relevance. The level scale used to define medical relevance was: *very important, fairly important, important,* and *slightly important.* Table 2 presents the weight values of the clinical section types used in the experimental evaluation, grouped by level of medical relevance. Table 2 shows that the weight of the diagnosis section was defined with a high value of $W_{Diagnosis} = 0.16$. On contrary, the excuse notes section was defined with a lower weight of $W_{Excuses} = 0.04$.

Table 2. Weight of conceptual component types.

Very Important (Weight 0.16)	Fairly Important (Weight 0.12)	Important (Weight 0.08)	Slightly Important (Weight 0.04)
Diagnosis	Consultation reason	Medication	Message agents
	Current illness	Studies	Advisors
	Physical examination	Procedures	Excuse notes
		Referral	Observations

All weights presented in Table 2 influence the calculation of the similarity metric of the learning method. Equation (16) shows the consistency of presented weights used for the similarity metric.

$$\sum_{sec_1}^{sec_n} W_{sec} = \sum_{very\ important} W_{sec_v} + \sum_{fairly\ important} W_{sec_f} + \sum_{Important} W_{sec_i} + \sum_{slightly\ important} W_{sec_s}$$

$$\sum_{sec_1}^{sec_n} W_{sec} = 0.16 + 3 \cdot 0.12 + 4 \cdot 0.08 + 4 \cdot 0.04 = 1 \tag{16}$$

7.3. Performance Evaluation

An experimental evaluation was conducted in order to analyze the lazy nature of the proposed learning method. In the learning approach, a similarity metric between clinical CTs is calculated by using all previously recorded CTs as training examples. Since the problem-solving ability of the proposed method is increased with each newly defined CT, it is important to analyze the performance of the proposed learning method when faced with CKBs with a great number of CTs.

7.3.1. Execution Platform of Performance Evaluation

The execution time analysis was performed on an Intel(R) Core(TM) i7-4700MQ CPU @ 2.40 GHz, 16 GB RAM, and running 64-bit Windows 10 Pro.

7.3.2. Execution Time

The efficiency of the learning method was evaluated when faced with CKBs of different sizes. To make a realistic evaluation, the 250 CTs registered by students were considered as input data, and the average time of 50 executions was measured for each CKB analyzed.

Figure 4 presents the average execution time of the proposed method when using different CKB sizes. The algorithm was executed on CKBs containing from 25 to 250 CTs.

Figure 4 shows how the size of a CKB has a direct influence on the execution time of the proposed method. Results also demonstrate that the proposed learning method is able to process 250 CTs in less than 90 milliseconds.

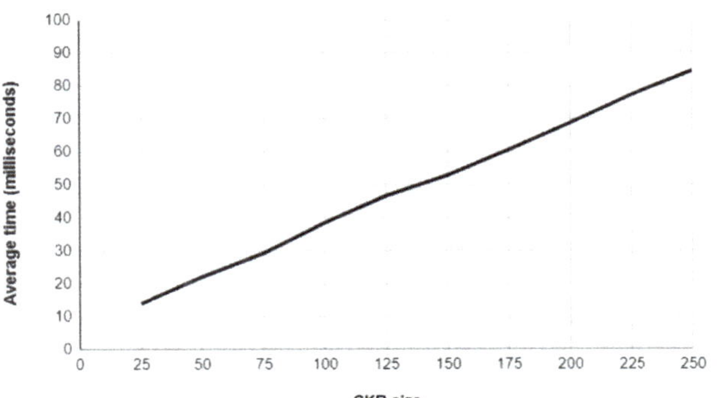

Figure 4. Average execution time of the proposed learning method regarding different CKB sizes.

7.3.3. Execution Time Projection

In order to estimate the efficiency of the learning method when facing larger CKBs, auxiliary CTs were generated based on the information recorded by the students. Although the auxiliary CTs were artificially built and do not reflect real clinical scenarios, they can be used to evaluate the performance of the learning method as they have the same data dimension as the CTs written by the students. The graphic in Figure 5 reports the average time of the proposed method regarding CKB sizes.

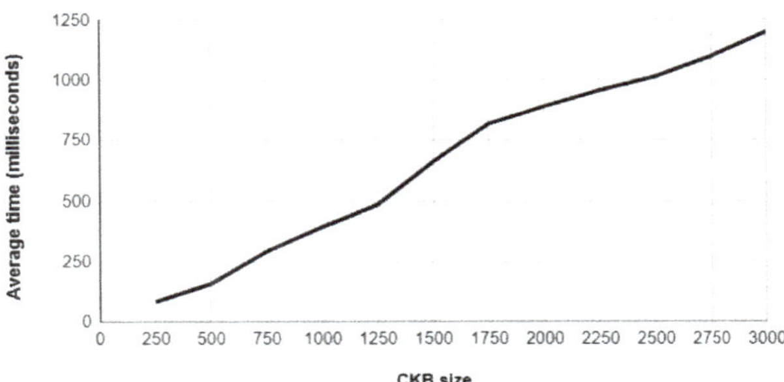

Figure 5. Average execution time of the proposed learning method when facing larger CKBs.

Figure 5 shows that the learning method generated suggestions in less than 1.25 s, even when facing larger CKBs with a great number of CTs. Given that 3000 represents a suitable bound for the number of CTs included in a physician CKB, the proposed method is able to produce suggestions in reasonable execution times, even when processing real CKBs with several workday scenarios.

7.3.4. Comparison with a Bayesian Learning Approach

To further analyze the applicability of the proposed approach, this subsection presents a comparison of the proposed instance-based learning method with a Bayesian learning method, which is based on a classical algorithm described by Mitchell [16] for classifying text documents.

The implemented Bayesian learning method works under the assumption that the occurrence probability of a word is independent of its position within a document. During the learning task, all medical records are examined as training examples, aiming at extracting the vocabulary of all words appearing in patient histories. After that, the frequency of each word is computed on all case types, to obtain the probability estimates of the Bayesian approach. Finally, to classify a new draft of the register editor, the probability estimates are used to determine the most likely case type to be applied.

Figure 6 reports the average execution time of both the Bayesian method and the instance-based method, regarding different CKB sizes, when processing the testbed of 250 CTs registered by students.

The graphic in Figure 6 uses a logarithmic scale to highlight the difference of two orders of magnitude between the execution times of both learning methods. The efficiency results reveal that the Bayesian learning method is difficult to apply due to high execution times, even when processing small volumes of data. Execution time results also reaffirm the benefits of the instance-based learning method, which significantly outperforms the Bayesian approach in terms of efficiency.

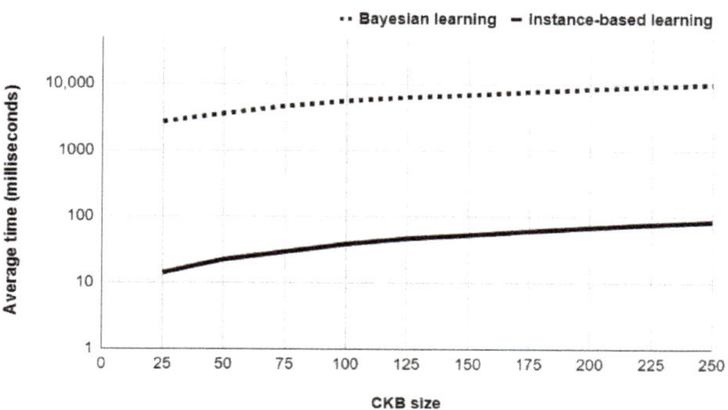

Figure 6. Average execution time comparison: instance-based learning vs. Bayesian learning method.

7.4. Testing the Applicability of the Instance-Based Learning Approach

In order to test the applicability of the proposed approach, a prototype was developed and deployed on Google Compute Engine, the Infrastructure as a Service component of the Google Cloud Platform. The prototype was evaluated by advanced medical students in their last year of training at Universidad de la República, Uruguay.

7.4.1. Comparison with Praxis

Praxis reports the average time required to write a CT starting from an empty CKB [15]. In order to compare the proposed approach with the original implementation of Praxis, the average time to write a CT using the prototype was measured. Figure 7 illustrates both average writing times starting from an empty CKB, by considering the medical attendance of the first 50 patients.

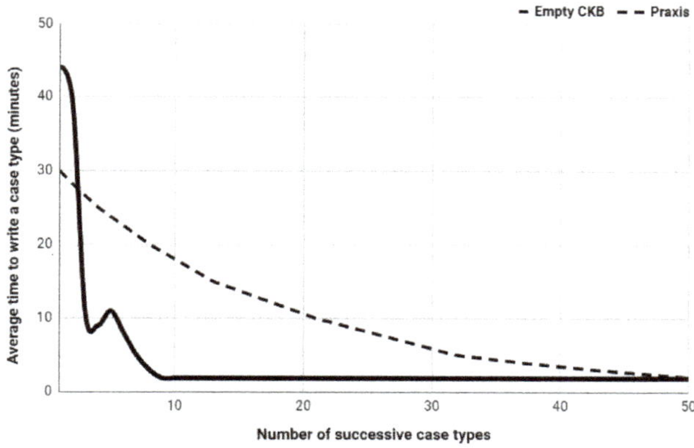

Figure 7. Average time of 50 medical students to write the notes of a case type (continuous line). Average time according to Praxis reports (dotted line). Both evaluations start with an empty CKB.

Although Praxis presents shorter registering times for the first two medical consultations, more than 50 evaluations are needed to achieve a convergence point. The proposed approach significantly reduces registration times from the third case registered onwards, converging quickly to less than three minutes of writing consultations. The proposed learning method demanded 210 min to register 50 consultations (i.e., 4.2 min per consultation), while using Praxis requires 519 min (10.4 min per consultation). The overall time reduction factor is $2.5\times$.

7.4.2. Improvement Using a Pre-Loaded CBK

The time needed to register a medical consultation can be reduced by using previously registered information. The average time to record a CT was measured in a context in which medical students could use a pre-loaded CKB. Figure 8 shows the average time to write a CT taking advantage of a pre-loaded CKB containing typical workday scenarios. As a relevant result, the use of a pre-loaded CKB implied a reduction of up to five minutes for recording the notes of the first six medical consultations. Furthermore, a pre-loaded CKB also accelerated the convergence to three minutes of writing consultations.

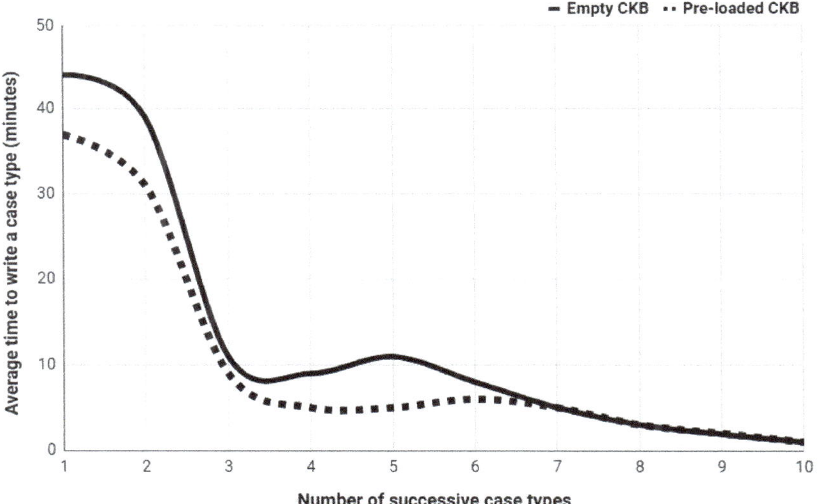

Figure 8. Average time of 50 medical students starting with an empty CKB (continuous line). Average time of 50 medical students taking advantage of a pre-loaded CKB (dotted line).

Regarding the scalability of the incremental processing of new case types, results suggest a convergence towards a short writing time for medical consultations, even when processing large volumes of data.

7.4.3. Survey about the Proposed Approach

More than 50 medical students from different editions of the Medical Informatics course were surveyed after using the prototype of the proposed approach. The advanced medical students have tested the prototype during course editions from 2016 to 2020. Figure 9 summarizes the best features identified by students.

Results show that 43% of the surveyed students mentioned that the learning curve was steep before they could benefit from the proposed learning method. As a relevant result, more than 73% of the students considered the prototype as an appropriate tool for medical practice, especially at medical consultations. Moreover, 62% of the students were able to speed up writing time during medical consultations.

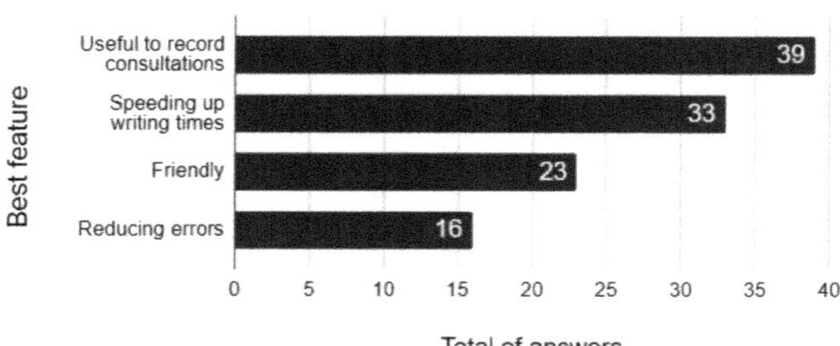

Figure 9. Best features of the proposed approach, according to the survey performed on students.

7.5. Interoperability of Health Information

Health terminological standards were taken into account due to the relevance of the interoperability of information in the Medical Informatics area. In particular, the national drug dictionary of Uruguay and a terminology server provided by the Hospital Italiano de Buenos Aires (HIBA) were integrated into the proposed approach.

7.5.1. Integrating the National Drug Dictionary of Uruguay

A National Drug Dictionary (DNMA) is defined by *Salud.uy* in order to standardize the information and vocabulary for clinical and logistical use applied to pharmaceutical and related products. DNMA acts as a standard of drug reference terminology for the network of healthcare service providers in Uruguay. Access permissions were requested from the DNMA in order to import the national dictionary of medicines into the proposed approach. Importing the national drug dictionary helped build a functional model, in which physicians can make pharmacological indications using a wide range of drugs.

7.5.2. Using the Terminology Server of Hospital Italiano de Buenos Aires

A terminology server allows linking the free text entered by a physician in a medical record to different health classifications, such as ICD9-CM, ICD10, or LOINC [4]. The use of a terminology server allows clinical information to be recorded in a structured form, using clinical terminology standards. Terminology standards enable interoperability of clinical information, and also allow information to be re-used for other purposes, such as clinical decision support, data analysis, and research.

The proposed system is able to use the terminology server supported by HIBA. The terminology server publishes its terminological terms grouped in different domains. This work has been successful in using terminology services for the domains that cover: reasons for consultation, diagnoses, procedures, and studies, which are required for the Uruguayan medical records model.

8. Discussion

The experimental evaluation of the proposed instance-based learning method focused on the practical aspect of the research. Thus, the evaluation was performed on a real use scenario, where the proposed approach demonstrated advantages over the original implementation of Praxis. Additionally, results were better in terms of writing times when using a pre-loaded CKB, containing typical workday clinical scenarios. Regarding the usability of the proposed system, a survey performed on a group of advanced medical students showed a high rate of approval. The implemented prototype was highlighted as an appropriate tool for medical practice and useful at medical consultations. Furthermore, and despite the lazy nature of the proposed method, the results showed that the learning approach was able to produce suggestions in reasonable execution times, even when dealing with large volumes of data.

Specific strategies can be applied to reduce uncertainties, including using expert knowledge to design and generate useful realistic instances for learning, and expanding the similarity metrics considered for the comparison of clinical information. In this regard, a recommendation for the practical aspect of the research is gathering and organizing as much information as possible about clinical practice in a systematic way, in order to help the automatic system to expand its base of knowledge to generate more accurate suggestions. In turn, a physician should be properly trained to register all the relevant data for the proposed learning-based system, without omitting any important information.

In this context, the main research lines for future work are related to evaluate the proposed system in a professional work environment of healthcare attention, with the aim of improving the accuracy of the learning method based on professional feedback. Thus, a future work line includes studying the proposed approach with the help of professional physicians. Another possibility for future work is related to enhance the accuracy of the learning method by improving the comparison between units of thought (clinical phrases). The weights of the clinical sections of case types used in the experimental analysis were defined simply, according to qualitative ranges of medical relevance. Consequently, a future work is to enhance the results by considering more accurate weights of medical relevance.

9. Conclusions

This work presented a novel approach to represent clinical knowledge, which supports an appropriate methodology for recording medical consultations. An instance-based learning method was also proposed, aiming at providing pertinent suggestions for physicians. Different scenarios of medical consultations were modeled to address the diversity of situations of physician workday, including multiple diagnoses and the attention of chronic patient. The proposed formal structure was also designed to use standard health terminology and codification. The approach was validated on a real case study involving 250 real instances constructed by advanced medical students. The proposed instance-based learning method was able to generate suggestions in reasonable execution times, even faced with large volumes of data. A total of 62% of the participants reduced the writing time of their medical consultations, which demonstrated that the approach was useful to accelerate the clinical registration process. Furthermore, results indicated it was appropriate to follow physician reasoning, especially during medical consultations. More than 73% of the participants approved a prototype following the proposed approach for assistance during consultations.

The proposed clinical representation supported by the learning method contributed to generate medical records faster than when using mainstream EMR systems. Overall, the proposed approach is a first step to explore new ways to foster physician thinking, overcoming difficulties of template-based clinical systems that are not designed from the medical point of view.

Author Contributions: Conceptualization, M.G. and F.S.; methodology, M.G.; software, M.G.; validation, M.G. and F.S.; formal analysis, M.G. and S.N.; investigation, M.G. and S.N.; resources, M.G. and F.S.; data curation, M.G.; writing—original draft preparation, M.G. and S.N.; writing—review and editing, S.N.; visualization, S.N.; supervision, F.S. and S.N.; project administration, M.G.; funding acquisition, M.G., F.S. and S.N. All authors have read and agreed to the published version of the manuscript.

Funding: This research was partially funded by ANNI, Uruguay and Infor-Med Corporation, Argentina. Part of the publication fee was funded by PEDECIBA, Uruguay. The work of S. Nesmachnow was partly funded by ANII and PEDECIBA, Uruguay.

Conflicts of Interest: The authors declare no conflict of interest.

References

1. Castellano, G.; Casalino, G. Special Issue "Computational Intelligence in Healthcare". *Electronics* **2020**, *9*. Available online: https://www.mdpi.com/journal/electronics/special_issues/CI_healthcare (accessed on 21 June 2021).

2. López-Rubio, E.; Elizondo, D.A.; Grootveld, M.; Jerez, J.M.; Luque-Baena, R.M. Computational Intelligence Techniques in Medicine. *Comput. Math. Methods Med.* **2015**, 1–2. [CrossRef] [PubMed]
3. Habib, J. EHRs, meaningful use, and a model EMR. *Drug Benefit Trends* **2010**, *22*, 99–101.
4. González, F.; Otero, C.; Luna, D. Terminology Services: Standard Terminologies to Control Health Vocabulary. *Yearb. Med. Inform.* **2018**, *27*, 227–233.
5. Infor-Med. Praxis Web. 2021. Available online: http://www.praxisemr.com (accessed on 1 April 2021).
6. Wang, Y.; Tian, Y.; Tian, L.; Qian, Y.; Li, J. An Electronic Medical Record System with Treatment Recommendations Based on Patient Similarity. *J. Med. Syst.* **2015**, *39*, 55. [CrossRef] [PubMed]
7. Klann, J.; Szolovits, P.; Downs, S.; Schadow, G. Decision support from local data: Creating adaptive order menus from past clinician behavior. *J. Biomed. Inform.* **2014**, *48*, 84–93. [CrossRef] [PubMed]
8. Nakai, T.; Takemura, T.; Sakurai, R.; Fujita, K.; Okamoto, K.; Kuroda, T. Prediction of Clinical Practices by Clinical Data of the Previous Day Using Linear Support Vector Machine. In *Innovation in Medicine and Healthcare 2015*; Springer: Cham, Switzerland, 2016; pp. 3–13.
9. Barbantan, I.; Porumb, M.; Lemnaru, C.; Potolea, R. Feature engineered relation extraction-Medical documents setting. *Int. J. Web Inf. Syst.* **2016**, *12*, 336–358. [CrossRef]
10. Shen, Y.; Colloc, J.; Jacquet-Andrieu, A.; Lei, K. Emerging medical informatics with case-based reasoning for aiding clinical decision in multi-agent system. *J. Biomed. Inform.* **2015**, *56*, 307–317. [CrossRef] [PubMed]
11. Installé, A.; Van Den Bosch, T.; De Moor, B.; Timmerman, D. Clinical data miner: An electronic case report form system with integrated data preprocessing and machine-learning libraries supporting clinical diagnostic model research. *J. Med. Internet Res.* **2014**, *16*, e28. [CrossRef] [PubMed]
12. Zieba, M. Service-Oriented Medical System for Supporting Decisions with Missing and Imbalanced Data. *IEEE J. Biomed. Health Inform.* **2014**, *18*, 1533–1540. [CrossRef] [PubMed]
13. Benmimoune, L.; Hajjam, A.; Ghodous, P.; Andres, E.; Talha, S.; Hajjam, M. Hybrid reasoning-based medical platform to assist clinicians in their clinical reasoning process. In Proceedings of the 6th International Conference on Information, Intelligence, Systems and Applications, Corfu, Greece, 6–8 July 2015; pp. 1–5.
14. Wilk, S.; Michalowski, M.; Michalowski, W.; Rosu, D.; Carrier, M.; Kezadri-Hamiaz, M. Comprehensive mitigation framework for concurrent application of multiple clinical practice guidelines. *J. Biomed. Inform.* **2017**, *66*, 52–71. [CrossRef] [PubMed]
15. Low, R. *The Theory of Praxis Concept Processing*; Technical Report; Infor-Med Corporation: Buenos Aires, Argentina, 2015.
16. Mitchell, T. Instance-base learning. In *Machine Learning*; McGraw-Hill: New York, NY, USA, 1997.
17. Sadegh-Zadeh, K. *Handbook of Analytic Philosophy of Medicine*; Springer: Berlin/Heisenberg, Germany, 2015.
18. Sociedad Andaluza de Medicina Familiar y Comunitaria. *Casos Clínicos en Atención Primaria*; Fundación Sociedad Andaluza de Medicina Familiar y Comunitaria: Granada, Spain, 2017.

MDPI
St. Alban-Anlage 66
4052 Basel
Switzerland
Tel. +41 61 683 77 34
Fax +41 61 302 89 18
www.mdpi.com

Applied Sciences Editorial Office
E-mail: applsci@mdpi.com
www.mdpi.com/journal/applsci